Sherman Adams

Homeland

A description of the climate, productions, resources, topography, soil,

opportunities, attractions, advantages, development and general

characteristics of Polk county, Florida

Sherman Adams

Homeland

A description of the climate, productions, resources, topography, soil, opportunities, attractions, advantages, development and general characteristics of Polk county, Florida

ISBN/EAN: 9783337289157

Printed in Europe, USA, Canada, Australia, Japan

Cover: Foto ©berggeist007 / pixelio.de

More available books at **www.hansebooks.com**

OF THE

CLIMATE, PRODUCTIONS, RESOURCES, TOPOGRAPHY, SOIL,
OPPORTUNITIES, ATTRACTIONS, ADVANTAGES, DEVEL-
OPMENT AND GENERAL CHARACTERISTICS

OF

POLK COUNTY,

FLORIDA.

By SHERMAN ADAMS.

1885.

TIGNER, TATUM & COMPANY,
BARTOW, FLORIDA.

PRINTED AT THE TIMES-UNION BOOK ROOMS, JACKSONVILLE, FLA

HOMELAND.

The desirability of a home in this fair, fertile, balmy, healthful and progressive interior section of South Florida, in Polk County, where cold and stormy winter never enters; where plenty abounds, enlivened with bright and soul-cheering sunshine, fragrant and healthful breezes, overflowing with the balsamic aroma of the majestic pine forests and the perfume of countless flowers; the attractions and advantages of a home in HOMELAND will be considered in these pages.

Here can be made the true HOMELAND, for which the human race has ever sought. Here can be realized humanity's brightest dreams of an Eden upon earth, with no tempter serpent to awaken unholy desires, nor flaming sword to prevent the full enjoyment of the choicest delights that can be appreciated by the most refined, virtuous, intelligent and contented people, that may here make for themselves the most pleasant and delightful of homes.

In HOMELAND every one can sit beneath the shade of his or her own vine, fig and orange tree, with none to molest or make afraid, while bounteous plenty, like a fair and loving goddess, pours the contents of her overflowing cornucopias at their feet; the choicest products of the temperate and semi-tropic zones.

Here gaunt and ravenous want is unknown; pinched and shivering poverty has no place; fell disease is shorn of its most virulent power; flowers bloom all the year; fruits and vegetables ripen at all seasons; the feathered songsters make the air melodious with their tuneful notes of joy, love and praise; bright, sunshiny days impart bouyant health, strength and cheerful thoughts; delicious breezes, impregnated with the balsamic healing of the pine, and the quickening and inspiriting odors of delicious flowers refresh and soothe; the cool, tranquil softness of the nights invites to calm, grateful, restful, life-renewing repose; the days are warm, but tempered by fresh and genial breezes; while the mornings and evenings are inexpressibly delicious, calm and mild, possessing a rare inherent charm, that once experienced is never forgotten, though it cannot be adequately described.

POLK COUNTY.

THE eyes of the whole world are directed towards Florida as never before, the interest growing rapidly with each succeeding year. Tired of cold, inclement winter, tired of the hard struggle to keep the wolf from the door, disappointed in hopes and expectations of securing either a competence or wealth, wearied in mind by continuous exhaustive effort, imbued with a strong desire to secure a fortune, or, enfeebled in body by the diseases so prevalent in various sections of the country, people everywhere are desirous to make a change for the better.

Reports, faint and few at first, and more than counterbalanced by those of an opposite character, have, for the past few years, been heard from Florida, from the land where, more than three centuries ago, notable Spaniards sought health—a fountain of eternal youth—and others sought wealth—the glistening yellow gold. Both are to be secured in this fair and balmy land, if rightly sought. The truth of the message, "Seek and ye shall find," can here be realized by all truly earnest souls.

The better reports have grown stronger and more numerous from year to year, until at the present time there are few sections of the Union that have no representatives in lately despised and bitterly-maligned Florida. Thousands have here secured restored health, and other thousands have gained fortunes, or at least a satisfying competence, as well as health.

Consequently, diligent inquiry and earnest effort is being made by the more energetic people of all classes, in all sections, to learn the exact facts with regard to this delectable land of sunshine and balmy breezes. "Are the good reports true regarding Florida?" is the earnest and oft-repeated inquiry.

This question, and many others, it will be the endeavor of this pamphlet to answer as regards Polk County, an interior section of the delightful South Florida peninsula that its residents, as well as the rapidly increasing number of visitors, consider the very choicest and most desirable locality that the whole world affords. Its lands are fertile, its waters are pure and abundant, it is noted for its healthfulness, its climate is the most delicious that can anywhere be found, its surface is diversified with hill and dale, meadow and plain, numerous lovely clearwater lakes reflect the bright sunshine, while their banks afford delightful building sites; here are running streams of sparkling water; here you can listen to the mocking bird to your heart's content, or watch them in their sportive glee; sturdy oak as well as beautiful pine and other trees abound on the uplands, while along the courses of the numerous streams are hickories, oak of several varieties, maple, gum, cypress, wild orange, cabbage palm, whitewood, magnolia and other varieties too numerous to mention. The surface of hill, plain and valley is covered with vigorous luxuriant native grasses, as well as beautiful trees; the numerous lakes and streams are well stocked with fish, turtle, etc.; cattle, swine and sheep thrive on its fertile ranges, in winter as well as in summer; sheep raising is found to be very profitable; poultry thrive and give quick returns; honey is abundant, bees increasing rapidly and laying up large stores; the growing of corn is more profitable than in the far West; sweet potatoes, pease, rice, sugar-cane, cassava, etc., are staple crops; all the citrus and many other semi-tropical fruits grow vigorously and produce abundantly; garden vegetables give immense returns; all kinds of industries thrive; new settlers are locating rapidly; the residents are eminently sociable and hospitable, kindly welcoming new comers; numerous business and social centres are already established, with churches, schools, stores, etc.; well regulated railroads give close and continuous connection with the roads of the whole country; other roads will soon be built, and the present roads be further extended; desirable locations abound, and lands can now be purchased at moderate prices, but are rapidly increasing in value; the roads are hard and firm, as well as free from mud or dust; snow is unknown and frost is rare; the country is exempt from cyclones, tornadoes, electrical storms and destructive winds, it being in a neutral zone, where it is impossible for such storms to enter. Many other pleasing and attractive characteristics might be mentioned, but I desire to be brief.

LOCATION OF POLK COUNTY.

" But," says the reader, "where is this wonderful land, this healthful, productive, attractive and charming County of Polk?"

It is located near the centre of the Florida peninsula, which is destined to be the most populous and wealthy, as well as the most enjoyable and desirable part of the known world. It lies mostly between the twenty-seventh and the twenty-eighth parallels of north latitude, and east of the eighty-second degree of longitude, an area that Judge J. G. Knapp, a prominent and well informed writer on Florida, designates as the Central Zone. It is, in fact, the "golden mean" between extremes that is so sincerely desired and earnestly sought.

ELEVATION.

Polk County has no high mountains, but a broad plateau of high land extends through it, near the centre, from north to south, the road-bed of the South Florida Railroad at Haines City reaching an elevation of 210 feet, about the same elevation being maintained at Lakeland, some twenty-five miles to the west. Some of the hills are still higher, being from forty to sixty feet above the surface of some of the lakes.

FRESH WATER LAKES.

An incomparable charm is given to the surface of the country by the hundreds, if not thousands, of lovely clear water lakes, many of them with attractive shores that afford delightful residence sites, that are scattered promiscuously over the face of the county. They are the most numerous, however, just north and northeast of the centre of the county, forming the far-famed Lake Region of Polk. In size they differ greatly, varying from one acre to five thousand acres in extent. On the eastern border of the county are the great Lakes Tohopekaliga, Cypress and Kissimmee, with respective elevations of 64.5, 62 and 59.66 feet above the sea level. South of Lake Cypress, in Township 28, Range XXIX, is Lake Hatch-e-ne-haw, with an elevation of 60.23 feet. In Township 29, Lake Rosalie, in Townships 30 and 31, Lake Walk-in-the-Water, 61.94 feet above the sea. Also, Lake Arbuckle, in Township 32.

WATER-COURSES.

The Kissimmee River, a broad and navigable stream, forms the eastern boundary of the county, separating it from Brevard and connecting the three great lakes first mentioned. Arbuckle Creek connects a series of lakes to the west. By cutting connecting channels the Drainage Company has made navigation practicable through lakes and river to the Caloosahatchie River, and through that to the Gulf of Mexico. The result has also been the drainage of large quantities of overflowed lands. In fact, the configuration of Polk County is such, and its elevation so considerable, that portions needing it can be easily drained, it being highest in the centre.

Peace River, which has its source in Lake Hamilton and its tributaries in the north-eastern portion of the county, Range XXVII, runs in a southwest course to Range XXV, Township 29, where it is augmented by the waters from Lake Hancock, flowing through Saddle Creek. It then continues southward in Range XXV. It could be made navigable the whole distance at a moderate expense, and undoubtedly will be in the near future, a charter having been secured for the purpose.

Tiger Creek and Arbuckle River, in the eastern part of Polk County, connect the series of large lakes found there with the Kissimmee River, and are being made navigable, which will give excellent facilities in the matter of transportation, and aid greatly in the development of that attractive section.

The Alafia and Hillsboro' Rivers and their branches intersect and drain the fine lands in the western part of the county. It was among the headwaters of these rivers that the first settlements were made in the county some forty years since, Tampa, in Hillsboro' County being the seaport and trading headquarters. Polk County is well watered by numerous creeks and streams of pure running water.

THE GOVERNMENT SURVEY.

Referring to the survey made by the General Government, it will be seen that Polk County extends north and south through Townships 25 to 32, inclusive. At the south it is included in Ranges XXIII to XXXI, while the northwestern corner that projects up to the western border of Orange and into Sumter County is included in Ranges XXII to XXVI. The area of the county is 2,060 square miles or 1,388,400 acres.

CONFIGURATION OF THE COUNTRY.

A comprehensive, or bird's-eye, view of Polk County shows it to be divided by nature into strips of a few miles wide, extending in a northerly and southerly direction. Near the centre, from east to west, is evidently the finest, most attractive and productive strip of country; at all events, here are the most improvements and much the greater portion of the population.

Lakeland, Acton, Auburndale and Sanataria, on the South Florida Railroad, some fourteen to sixteen miles in a northerly direction from Bartow, the county seat, are active and thriving new towns, the growth of the past year. To the north of them for a number of miles is a fine high country that will soon be developed, as the Florida Southern Railroad passes through it from north to south, forming a junction with the South Florida at Lakeland. To the South, fourteen miles to Bartow, six to the Bethel neighborhood, six more to Fort Meade, and a few miles beyond to the southern boundary, is a very notable strip of high pine and oak land from three to five miles wide. On this are many clearings, cultivated fields, fruitful groves, pleasant and even elegant residences, and three busy, rapidly-growing towns, Lakeland, Bartow and Fort Meade, with churches, schools, hotels, stores of varied kinds, public halls, post-offices, express and telegraph offices. The first two have railroad depots and the other doubtless will have in a short time.

This central, well developed strip of country lies to the west of Peace River. To the east is another apparently equally choice strip of land extending through the county ; but as yet, however, thinly populated, there being neither post-office nor store. At short intervals on either side of Peace River are creeks, or rivulets, that empty into it the drainage of these fertile side lands, hammock at first, but gradually rising to productive oak and pine lands, then less fertile pine edged by a strip of varying width of a flat-woods character, interspersed with bays, cypress swamps, grass ponds, etc.

The northwestern portion of the county may be considered as a separate division by itself, as Judge Knapp does with the northwestern part of the State. Its characteristics are those of Hernando and Sumter rather than of Polk proper, the bulk of which lies below the 28° of latitude, while this projection of four ranges, between four and five townships in depth, is north of that line. Consequently, it is in the North Central, instead of the Central Zone, according to the very convenient division of the State of Florida by Judge J. G. Knapp into the Northwestern, the Northern, the North Central, the CENTRAL (in the northern part of which the most of Polk county is situated), the South Central, the Southern, the Semi-Tropical and the Tropical Zones, each of which, with the exception of those at the extreme north and south of the State, occupy a full degree of latitude.

This "baker's dozen" of townships contains considerable quantities of excellent land. There are also many ponds, cypress swamps, flat lands, etc., especially on the eastern half adjoining Orange county, which, by some freak of the law-makers here, adds a half dozen townships to its southwestern portion. It is certainly high time for a change in the State Constitution, that the boundaries of a number of the counties may be revised.

Along the western border of the county, expecially in the south-western portion, and also along the eastern, the lands are of a flat-woods character, a rim of which virtually extends around the county. The soil is mostly fertile and there are numerous knolls, or broken ridges, that supply excellent sites for buildings, groves and cultivated fields, while the lower lands afford nutritious grazing for fine herds of cattle, swine, etc. These opportunities are being improved and the residences of farmers and cattle-men are well scattered over the whole county, especially in the western section.

THE LAKE REGION.

North of Bartow, in Range XXV, Township 27, the beautiful Lake Region proper commences and extends in an east and southeast direction through Polk to the centre of the eastern portion of Manatee County, which is situated in the southern part of the delightful Central Zone. There is also a group of fine lakes in Range XXIV, Townships 27 to 29. The drainage of the whole eastern part of the county tends to the great Lake Okeechobee, from which the Drainage Company is cutting canals to Gulf and Ocean, thus insuring the drainage of an extensive tract of country.

That which is designated as the Lake Region proper lies to the northeast of Bartow, beautiful sheets of water being here grouped very thickly together. The South Florida Railroad passes through the heart of the lake system, and new towns are springing up as if by magic. The more important thus far are Haines City, Bartow Junction and Winter Haven, which, though of the present year's growth, are developing very rapidly, the chief attractions being the rare beauty and healthful salubrity of the country. The soil, though not as fertile as in the exceptionally fine strips of productive country on either side of Peace River from Bartow to Fort Meade, which is the most desirable that can be found in Florida, is nevertheless well adapted to citrus fruits. Along the margins of the lakes and small streams are many tracts of very fertile land well adapted to the profitable production of all kinds of vegetables for market, the numerous sheets of water being a good guarantee of protection against frost. Strawberries, pine-apples, bananas, etc., also succeed well in these sheltered localities. Judging from present indications, but a short time will pass before this whole lovely region will be very thickly populated.

LAKE HAMILTON TO LAKE ARBUCKLE.

Going southeastward from Haines City, we find a high ridge of land several miles in width, with lovely lakes to the east and to the west. This has been so unfavorably situated as regards transportation that almost the entire surface is yet covered with the primeval forest. Settlers are pouring in there, however; railroads will soon follow, and an entire transformation will be wrought in a very few years, while the hardy pioneers will have secured fortunes. Even now a railroad is projected from Haines City to Rosalie, a new and vigorous town on the lake of the same name. Steamers run to this place from Kissimmee, on Lake Tahopekaliga, and past here through the Drainage

Company's canals that connect the several lakes with the Caloosa-hatchie River and the Gulf of Mexico. Those who have traveled through the section, from Lake Hamilton on the north to Lake Arbuckle at the southeast, are enraptured with the beauties of the high rolling country and the many advantages to be secured by the settler. There is little opportunity to obtain homesteads, as all the choicest tracts are already purchased or occupied.

The flat lands between this delightful ridge and the Kissimmee River are excellent for grazing, and when the Drainage Company shall have completed their operations, here will be found extensive and profitable fields of sugar cane, rice, etc. The cane fields at Rosalie are already giving wonderful returns.

GREAT VARIETY.

Unquestionably Polk County contains a greater variety of soil and scenery than any other section of the State, thus affording something almost sure to suit all tastes and desires, however varied. The diversity of vegetation is also very great, while the cultivatable crops embrace nearly every variety grown by civilized man in both the eastern and western hemispheres. The capabilities of the county are immense and the value of its products can be readily increased at least a thousand-fold by earnest and active intelligence.

DOMESTIC ANIMALS.

The chief source of wealth in the past, and an important industry at the present, is the raising of cattle for market, most of the business and professional men, except the later arrivals, having been "cowboys" in their youth, their herds grazing not only in Polk County, but also far to the southward. This industry is now somewhat depressed in consequence of the closing of the Cuban market, from whence they have drawn large store of gold ; but the rapid increase of emigration into Polk County will afford some relief by the increased consumption here.

PONIES are raised in moderate numbers, but the supply falls far short of the demand. Consequently the larger number of horses and mules are imported from other States.

SHEEP are not raised extensively as yet, but have been found to do well, giving excellent and profitable returns.

SWINE here find their paradise, the abundant "mast" furnished by the frequent oaks and the great quantities of esculent roots along the water courses giving them abundant food with little trouble.

POULTRY thrive here as in no other section of the Union. Broods of chickens are hatched every month in the year and grow rapidly, giving abundant returns.

AGRICULTURAL PRODUCTS.

CORN is an important crop and large quantities are raised for home use. Its growth secures to the cultivator more profit to the acre than is realized by the grower in the far West. The average yield is from ten to forty bushels per acre, according to the quality of the land.

FIELD PEASE are an easily-grown and profitable crop, affording a large amount of sustenance for both man and beast. The seeds are especially excellent for poultry and the vines are greedily eaten by cattle.

RICE yields bountifully both upon the uplands and the lowlands. The straw makes excellent forage.

SWEET POTATOES are a standard crop and always find a ready market at good prices. The yield varies with the quality of the land, the preparation and the attention given, ranging from one hundred to five hundred bushels per acre.

CASSAVA is also a very desirable and profitable crop, easily raised, and should be grown extensively.

SUGAR CANE gives excellent results. Fine fields of cane are quite common and profitable.

COTTON of fine quality has been raised, the soil being adapted to its growth, but other products can be grown with so much more ease that it receives but little attention.

WHEAT, barley, buckwheat and some other grains are but little cultivated, though there seems to be no known reason why they would not do well if planted at the right time and given proper care.

OATS have given excellent results at times, but are mostly grown as a forage crop and fed in the sheaf.

RYE is attracting attention as a soiling crop, and can be grown extensively with profit. Sown in the fall, it will grow all winter, giving a very pleasant appearance to the fields as well as profit to the owner.

THE PEANUT, or Pindar, here finds soil and climatic conditions very favorable and yields large returns.

CHUFAS, a species of ground-nut, are very productive and are excellent for promoting the growth and fattening of swine and poultry, who prize them highly and will help themselves whenever the opportunity offers.

FIELD BEANS are recommended as a sure crop, by high authority, if planted in June.

TOBACCO grows finely, but its culture is not advised, as it is a very exhaustive crop.

EXPERIMENTS should be made carefully and continuously with all known products. Some of the results will be agreeably surprising and profitable.

THE GRASSES.

The natural grasses are so abundant and some of them are so nutritious that but little attention has been given to the cultivated. Crab grass springs up and grows luxuriantly in cultivated fields, and ought to be utilized for hay. A kind of blanket grass makes excellent pasturage. The same is true of smut grass. Bermuda grass grows luxuriantly, and a mixture of this with smut grass would undoubtedly make excellent pasturage. I have also seen fine specimens of red clover growing in a number of places. It might thrive on the firm lands of Polk County. Alfalfa ought to be given a thorough trial throughout the county, as it is a great favorite in California and is winning high esteem in the Southern States. It might be grown in the orange groves, as its roots penetrate the subsoil; hence it would not be open to the objections urged against grasses that are surface feeders. St. Augustine grass has been highly recommended for lawns. The country seems naturally adapted to grasses, most of which, however, are too wiry when mature for fodder. There is no doubt but that by judicious care and attention good stands of cultivated grasses can be secured, and dairies be made numerous and profitable with as little trouble as in any part of the country. Every family can keep a cow, and there are no long cold winters to necessitate filling large barns with hay. They can be fed profitably with the large variety of green forage and root crops that here grow luxuriantly.

GARDEN VEGETABLES.

IRISH POTATOES give good returns. They are planted in the fall or winter and dug in the spring.

SWEET CORN is a profitable crop, for which there is a good demand for home use as well as for shipment. Those who desire can have it on their tables from Christmas to the 4th of July.

MELONS, squashes, pumpkins, etc., grow to large size and produce abundantly. In fact, Polk County seems to be the native habitat for vines, as all kinds grow luxuriantly.

CUCUMBERS, beans, tomatoes and cabbage are standard crops for shipment, and give large and profitable returns.

CABBAGE, cauliflower, and the like, find soil and climate especially adapted to their vigorous growth.

BEETS, turnips, carrots, parsnips, radishes, etc., here find favorable and satisfactory conditions, and yield abundantly.

EGG PLANTS, okra, lettuce, etc., do finely.

ONIONS grow to a large size and are of excellent flavor.

PEPPER PLANTS grow to the size of small trees and yield abundantly from year to year, at all seasons, in winter as well as summer. They are ornamental as well as profitable.

HERBS and plants for seasoning, as well as for medicinal uses, yield a supply for all needs with very little care and attention.

GARDEN VEGETABLES, with hardly an exception, give excellent returns when planted at the proper season. They, as well as other plants, are benefited by watering, in the event of a drouth, which sometimes prevails in April or May.

VARIED FRUITS.

STRAWBERRIES are a luscious and delightful fruit, as well as profitable. They are peculiarly adapted to the soil and climate of Polk County, and ripe berries can be had every week from December to June. They are always in demand here, but can be shipped North at a time to secure the very highest prices.

THE FIG does well and yields a good amount of pleasant and nutritious fruit

GRAPES grow wild in the hammocks, and may be cultivated with a good degree of success.

BANANAS grow vigorously and fruit well, the stalks attaining three feet to forty inches in circumference and twenty-five feet in height, with leaves from twenty to twenty-eight inches across and five to six feet in length.

THE GUAVA is a favorite fruit that grows well and fruits abundantly in most localities, but, like the banana and pine-apple, is quite susceptible to the effects of frost.

THE LIME is an important fruit for extensive cultivation, but, like the guava, requires favorable localities as regards exemption from frost. It has been recommended for hedges as well as for fruit.

PINE-APPLES may be profitably grown under the same conditions as the banana and guava, as regards exemption from frost, and will give very profitable returns. If grown in an exposed locality it would pay to give them protection in the event of probable frost, for there is no such thing as a frost line in Florida, though the low latitude of Polk County and the large numbers of lakes to the northward give the main body of the county exceptionably favorable conditions, superior to more northern localities; yet, even here, much depends on the situation, which can only be learned by personal observation and experience.

THE LEMON grows well, and is destined to be a very profitable fruit. It is more hardy than the guava and the lime, but less so than the citron, grape-fruit or orange. The genuine Sicily is the variety preferred.

THE CITRON, of which there are many varieties, has thus far been grown only for ornament, the proper mode of preparation for market being unknown; but that difficulty is about being overcome, and its cultivation will no doubt be very profitable.

THE GRAPE-FRUIT is the favorite for the spring-time, its extreme juiciness and sub-acid flavor making it very palatable and refreshing, as well as healthful.

THE JAPAN PLUM and persimmon are destined to be important fruits, but their culture is yet in its infancy.

THE PEEN-To and the Honey Peaches will no doubt become standard fruits but attempts at cultivation are very recent.

THE MULBERRY is of quick growth, makes a fine tree and yields an abundance of wholesome fruit.

THE CASTOR BEAN, or Palma Christi, here grows to the size of a tree, and yields abundantly year after year. The making of castor oil promises to become a profitable industry.

EXPERIMENTS are being made with a great variety of desirable fruits, and there is no doubt but that within a few years the list will be greatly extended.

THE ORANGE, however, is the king of all the fruits, the standard of excellence and chief dependence. All other fruits are merely accessories, side issues, at present, though some of them may eventually rival it in profitableness, if not in lasting durability. The fertile soil and delicious climate of Polk County combine to produce the most vigorous and fruitful trees and the most luscious fruit that can be produced in any part of the world, and that, too, with the most ease and rapidity, and at the least expense. This is due to the excellent quality of the soil, it requiring very little, if any, fertilizing, and the very mild and very short winters, thus giving nearly all the year for growth, which are pertinent facts well worthy of consideration. The trees here attain an immense size, the older ones yielding from 1,000.

to 10,000 each of the golden fruit. There being no destructive freezes here the beautiful and luscious fruit can remain on the trees all winter, if desired, and sold upon the most favorable market.

HOW TO MAKE A GROVE.

Directions for the making of a grove of orange or other fruits, or for the cultivation of vegetables or farm crops, the management of poultry, cattle, etc., have no place in a work of this character, the sole aim of which is to show what Polk County is, what has been and may be accomplished there, and the advantages it offers to the immigrant in the way of soil and climate, health and fortune. Instructions as to how work should be done are useless, until one is on the ground, ready to go to work.

The Agricultural Department at Washington has published instructions as to the making of a grove and other matters pertaining to Florida. Rev. T. W. Moore, of Fruit Cove, Fla., has published a standard treatise on orange culture, which, as well as several other agricultural works pertaining to Florida, can be procured of any book-seller or news-dealer.

To get at the true inwardness of a State, and especially of Florida, an acquaintance with its leading newspapers is indispensable. Every one desiring to know of Florida, should, as a first step, send $1.25 to the *Times-Union* office, at Jacksonville, Fla., C. H. Jones & Brother, publishers, and secure for that sum the *Weekly Times*, a large folio of thirty-six columns, and *Munroe's Annual*, an octavo pamphlet of about 300 pages, which contains an immense amount of statistical and other desirable information about Florida. The local papers of the section, in which you think you might be interested, would also be good investments. Their prices are from $1.50 to $2 a year. By getting the *Weekly Times* you secure not only an excellent family newspaper, but also matters of news and correspondence from all parts of the State, ably edited and well selected. *The Dispatch*, of Jacksonville, and *The Agriculturist*, of DeLand, are weekly papers that you will need after you get here to teach you what to grow and the best methods. *The Weekly Times* also has excellent articles in this line by J. G. Knapp, the experienced agricultural editor.

FIBRE PLANTS.

The tendency of the soil and climate of South Florida seems to be toward the production of fibre and but few years need elapse before the millions of dollars annually sent to the East Indies for fibrous materials can be kept at home to increase the wealth of the country. Jute, hemp and ramie, and a variety of other fibrous plants grow here with great vigor. Even the grasses here tend to fibre. Notably among them is found a plant growing wild in the woods, known as bear grass. It attains a length of three feet, and has a white fibre of wonderful strength. Jute is indigenous to the State, growing wild and becoming a pest or weed about the farms, as it springs up perennially. The saw-palmetto, with which thousands of acres are covered, is also a very valuable plant for fibre. From it are made brushes, mattresses, paper, etc.; but it is needless to continue the list. The fibrous productions of Polk County only await utilization by intelligent and enterprising men to develop immense wealth.

ROSES AND OTHER PLANTS.

The low latitude, the equable temperature, and the absence of destructive freezes, enable those who will to have their yards and gardens filled with flowers throughout the year, while their residences arce'mbowered in beautitul running vines. It also gives profitable opportunity to raise roses and other plants for Northern markets, where they bring excellent prices. There will also be an active home demand for flowers from the thousands of winter visitors. Those who have a taste for plant culture have here the source of a handsome income.

IVEYS, honeysuckles, Spanish goose-berry, cypress and a great variety of other vines thrive wonderfully.

INSECTS AND REPTILES.

None except those resident here have any idea of the bitter injustice that has been done to Florida, and especially to Polk County, as regards annoying insects and poisonous or dangerous snakes and other reptiles. The facts are that no part of the United States, or of America, in fact, is more free from pests of this character. As regards mosquitoes, they are so few that mosquito nets are unused, and unseen in a large part of the county. The same is true of sand-flies. Houseflies, too, are much less abundant than at the North. Fleas breed on hogs, but are not especially annoying after the first year. Roaches are no more common than in other parts of the South, and some places at the North. The quantity depends upon the neatness, or reverse, of the housewife. Gnats are no more troublesome or abundant than in other localities. The same may be said of the varieties of flies and other insects that are found in woods and fields all over the world. There are as few in Polk County as anywhere.

Poisonous and other snakes may be dismissed with a word. There are few, very few of them; less probably than in most sections of the Union.

'Gators have been hunted so extensively for their hides and teeth that they are becoming not only scarce, but timid, and keep at a safe distance.

GAME, FISH, ETC.

Game always disappears with the advent of civilized man. Polk County affords no exception. Deer, wild turkey, etc., have been plentiful. There are some foxes and squirrels, abundance of rabbits, coon, opossum, etc.; also quail and other game birds.

The lakes and streams are well stocked with black bass, catfish, bream, perch, soft-shelled turtle, etc., but people here fish and hunt for the sport, and not as a means of livelihood.

HONEY BEES.

Bees do extremely well in Polk County, and those who have a taste for apiculture can secure quite a revenue from this source. Florida honey is not only equal but superior to that of any other section. In this regard, as was proved at the New Orleans Exposition, even California has to take second place.

WHAT INDUSTRIES.

That Polk County has the capacity and the requisites for the successful prosecution of any and all the industries common at the North and West, except that of mining, will be self-evident to all who have perused the foregoing pages, besides a number peculiar to the country. It is also highly probable, so much so as to be a matter of almost absolute certainty, that the thousands of active men who are coming hither from all sections will originate many new industries, or adapt old ones to the needs of this section. Here are thousands of opportunities for earnest, clear-headed men to achieve fortunes.

THE SOIL.

That Polk County has a first-class reputation for excellence of soil has never been denied. In fact, it is credited with the possession of the best and most productive soil in the State. It also has the greatest variety, though the better class predominates. It has high and low, gray and black hammocks, poor and rich pine lands, productive oak lands, and barren scrubs, dwarf pine and black-jack ridges, wire-grass and saw palmetto lands, bay-gall and sand-flats, open prairie and grass ponds, rich bay-heads and cypress swamps and lakes of every conceivable size and variety of beauty. Every taste, desire and requirement can be gratified.

PRICES OF LANDS.

Prices are rapidly advancing, but they are so variable, and depend upon so many conditions, that it is virtually impossible to give any satisfactory idea regarding them. They range from $1.25 to $2,000 per acre, and depend upon quality, location—present and prospective, as regards business centres and railroads—and the views and necessities of the owner. Prices are rapidly changing, but the new price is invariably an advance on the previous price. Every lot cleared, every new house built, every railroad constructed, or new industry started, increases the selling value of all the lands in the neighborhood. Fortunes are being made in lands. Average prices range from $5 to $50 per acre.

CAUSES OR BASIS OF PROGRESS.

One important cause of the rapid growth and development of Polk County, since the South Florida Railroad was constructed through it, is due to its exceptionably favorable climatic conditions. The cheapness of the lands; their unusual fertility ; the great variety of productions of which they are capable ; the ease of obtaining a livelihood ; the unexcelled beauty of the country ; the many lovely lakes and numerous running streams; the excellence of the water ; the comparative freedom from insect pests and dangerous reptiles ; the remarkable healthfulness of the country ; the attractiveness of the abundant oak growths, reminding prospectors of more northern States ;

the kindly social and neighborly character of the people; the surprisingly firm character of the ground in the better portions; the absence of deep sands, insuring good roads and easy travel ; all these, and a large number of other reasons that could be adduced, give a great impulse to immigration, and the purchase of land as soon as the facts become known to the outside world ; but, thus far, Polk County has made but little show in newspapers, or in the pamphlets of advertising agents. The year and a half since the railroad reached her boundaries, or more properly the half year since the South Florida Railroad penetrated to the centre of the county, to Bartow, its county seat, has not given time for any extensive advertising. It is, however, developing very rapidly, because of its intrinsic merits.

CLIMATIC CONDITIONS.

Its low latitude, between the twenty-seventh and twenty-eighth degrees, and moderate elevation, as well as its numerous modifying and protective bodies of water, insure Polk County against destructive freezes, and give to its vegetation and products a semi-tropical character, as well as long seasons for growth, with very short and very mild winters.

Being not only located in the centre of a peninsula, but also within the region of the trade winds, that blow with unfailing regularity, it is sure of refreshing daily breezes, that both cool and purify the atmosphere, and make stagnant and sultry air absolutely impossible. Being insular between the broad Atlantic and the Gulf of Mexico, with the intervening spaces filled with balsamic pine forests, the air, as it filtrates through them, becomes heavily charged with their healing and health-giving aroma, in addition to the life-giving ozone from the Ocean, as well as with the perfume of countless flowers, more potent medicines than any physician can give.

The days are shorter and the nights are longer in summer than in sections further north ; hence, the earth has less time to become heated and more time to cool in summer than in higher latitudes. In addition to this, during the hot days of summer the evaporation is very rapid from lake and river, Gulf and Ocean, rendering much of the heat latent. The vapor rising, forms clouds which intercept the heat of the sun, shielding earth and man from its calorific rays. The moisture becomes excessive, and it falls in refreshing showers, cooling the atmosphere and preventing dust and drouth, as well as absorbing heat by speedy evaporation. Much of the heat is also dissipated by the breeze from the Ocean and borne across the narrow peninsula, here only a hundred miles wide, to the Gulf. Thus these several causes, and there may be others, combine to make the heat less, as well as the air more pure and strengthening, than at any distance to the north. Therefore, we find that the farther south we go on the Florida peninsula the less the altitude of the thermometer in summer, while the heat really felt is actually several degrees less than indicated. Thus, a really hot day is less exhaustive and more enjoyable than in any other part of the country. People generally do not understand these facts. When they do, South Florida will be a popular summer, as well as winter, resort.

In summer the ocean is cooler than the land, consequently breezes from the ocean, like the trade winds, make excessive heat impossible. Yet these trade winds extend but a short distance beyond the tropic circles, as will be seen by examination of any physical geography. As regards Florida, they are felt regularly only in the southern portion, in South Florida, and the farther south the greater their power. Their effects are also felt more completely on an elevated table-land, or plateau, like Polk County, which reaches its greatest altitude in the wide central strip or ridge that extends through the county from northwest to southeast, giving full scope to the winds and affording excellent natural, and opportunities for artificial, drainage. These facts make a residence in Polk County much more desirable than in the lower and flatter lands, by which this central ridge is surrounded on all sides for many miles. It is also preferable to, and more to be desired, than localities further north, both because of the superior benefit it gets from the trade winds, that give such excellent results in Polk County, purifying the air and mitigating the summer heats until they become very enjoyable, and the comparative absence of frost.

In winter the waters of the Ocean are warmer than the land. Hence, the effect of the breeze from the Ocean is reversed and winds from Ocean or Gulf are warm and enjoyable. These winds also pass over the warm Gulf Stream. It is only the northerly winds and those from the home of storms in the northwest, among the Rocky Mountains, that bring disagreeable cold. But these winds are modified and deflected, bent northeastward, by the prevalent winds from the east and south that sweep over the Gulf Stream's warm waters, which flow around the south end and up the east side of the peninsula.

Hence, though localities to the west, the northwest and the north, may suffer from disagreeable cold and even frost, the favored parts of Polk County are exempt, for several reasons, among which may be mentioned the fact of the lower latitude of Polk, and consequently greater natural warmth; the fact that the cold winds are beaten back by the prevailing winds, giving them a direction to the northeast; and, a final important consideration, the beneficial influence of all the lakes in the State to the north of Polk County, as well as the large numbers within her borders, in taking the frosty sting from winter's chilling winds. Any one can see that facts like these are self-evident. The lower the latitude, with the same or a less elevation, the higher the temperature in winter; the warmer and the more constant the winds from the east and the south, the less the cold that can reach the locality from the north, and the greater the number of lakes interposing, the more equable the temperature and the less the variation in the range of the thermometer. In this connection may be properly noted the fact that the days and nights are more nearly of equal length than in any locality at any distance further north. This, as will be readily seen, insures a greater time and amount of sunlight in winter and greater consequent warmth. Hence, we conclude that as regards mildness and equability of temperature, absence of cold, disagreeable winds and frost, with all the consequent advantages to be derived therefrom, Polk County stands without a peer—unequalled.

CYCLONES, TORNADOES, ETC.

Not only can Polk County justly claim a more genial, equable and desirable temperature than any other section of the State, greater freedom from insect pests, more varied lands and landscape, a more generally fertile soil, and a soil and climate that give opportunity for the profitable production of a greater variety of fruits and vegetables than any other section of the State, or of the Union, but it also has a great advantage over all other sections, except a moderate tract of interior country a few miles to the north and to the south of its borders, in the fact that it is situated in the centre of the narrow belt extending across the South Florida peninsula that is *exempt from*

DESTRUCTIVE STORMS,

as is proven by experience, and evidenced by careful and scientific study of the course which storms always take and the physical conformation of country that shapes the pathway, or route, of all severe storms. Neither cyclones, tornadoes, nor hurricanes, can ever travel over the fair surface of Polk County, leaving devastation and ruin in their track, as is so often the case in the West and Northwest, the North, and occasionally in the South. The scientific reasoning by which this is proven is rather abstruse and extended, and worthy of consideration. .We have not space for it in this work, as we are dealing only with facts, without extended reasoning as to the cause. The fact is patent that no such storm has ever visited this section, and science shows that it cannot. Let these facts be deeply pondered by those who live in those sections of the country where the cyclone, the tornado, the blizzard, the electrical storm, or even the fence-prostrating, chimney-tumbling, roof-lifting equinoctial storms have full sway. The people of Polk County are absolutely ignorant, as regards personal experience here, of the characteristics of a really severe storm. They do not know how to appreciate even a gale on the coast. One that has experienced "storms as are storms" cannot help smiling at the residents' relation of their experiences with, or in, storms that they considered severe. In the writer's four years' experience of South Florida he has not seen, known of, or felt any storm that can begin to be compared in severity with the usual equinoctial storms of the North or West, or with the gales that so often prevail on the eastern coast of any part of America. In this regard, Polk and Orange Counties are predominant. They have no equals anywhere. They are unexcelled, unapproachable, for in them the storm king is shorn of his power. How immense is the sense of security and enjoyment when you feel assured that your home is in a section where your crops will not be destroyed by unruly winds on a rampage; where your fences and barns will not be scattered over your fields, and where your house and your loved ones are secure.

EDUCATIONAL FACILITIES.

At present, the system of common schools supported by the State, and a few private schools, comprise the sum of all that is available in the matter of education. Great interest, however, is being awakened in the matter, excellent school buildings are being erected in several parts of the county, the very best of teachers are to be employed, and there is every reason to believe that the educational interests and facilities of Polk County will soon be fully equal to those of any county in the State, or in other parts of the country. The county already has a handsome school fund from the donations of Jacob Summerlin, which will doubtless be speedily increased.

RELIGIOUS INTERESTS.

The Baptists and Methodists have handsome churches and parsonages in several parts of the county, church societies are organizing, and these and other denominations will speedily erect other church edifices and supply them with an able ministry, in addition to the present regular preaching. Well-attended Sunday Schools are organized throughout the county.

MANUFACTORIES.

At present, saw and planing mills, with wood-working machinery for making pickets, laths, shingles, mouldings, etc., comprise the bulk of the county's manufactories. Bartow has a grist mill and a harness manufacturer. There has been a tannery and large boot and shoe manufactory at Fort Meade. There is also a brick-yard two and a half miles south of Bartow. Polk County offers rare opportunities for the establishment of a great variety of manufactories that would pay a very profitable percentage on the investments. Those who find business dull at the North can here retrieve their fortunes.

VARIETIES OF BUSINESS.

General merchandise stores take the lead, being established in all the more important places. The country is settling up rapidly, however; new centres are being established, and there are increasing opportunities for enterprising men with stocks of goods to build up a handsome business. Bartow and Lakeland have especial drug, hardware and some other stores, and the variety of business is rapidly increasing in those and other places. Most of the centres have railroad depots, telegraph and express, as well as post-offices. Hotels are numerous, and charges moderate. Most towns have one or two livery stables. There are, also, public halls and opera houses, skating rinks, millinery stores, soda and ice-cream rooms, billiard parlors, shoemaker's shops, news rooms, barber shops, photograph galleries, boarding houses, insurance and real estate agents, contractors and builders, attorneys, physicians, butchers, grain and feed stores and a variety of other occupations and industries. Last, but not least, Bartow, Lakeland and Fort Meade each have wide-awake newspapers—*The Informant, News*, and *Pioneer*—that put into print such matters as the respective editors deem of most interest and benefit to the country.

DESIRABLE INDUSTRIES.

The number of new and beneficial industries that might be established with profit to the proprietor and benefit to the communities can be counted by the tens and the scores. We will note but a few. Enterprising men, by a little reflection, can suggest many that would be likely to return a good income. Blacksmiths are needed in several places. Wagon-makers could find rapid sale for their products, as well as considerable business in the way of repairs. Good boat-builders are needed in the Lake Region. Machinist and repair shops with lathes and other desirable machines and tools would find a rapidly increasing amount of work. A few mills to saw out pickets, with machines to make an improved portable wire and picket fence, could do a lively business from the start; orange boxes and vegetable crates are in demand, and a surprising quantity would find a ready market; new furniture is in great demand, and there is plenty of excellent timber for its manufacture; a well-managed factory could do a large business. Many articles of domestic use are made of wood. Why not make them here, where woods are in great variety and suitable for nearly every conceivable purpose? Barrels are needed for sugar and syrup; a cooper could find constant employment. The wood-work, at least, of many agricultural and labor-saving implements, might be made here. Factories for the manufacture of the fibre of the saw-palmetto into material for mattresses and upholstery could here find abundance of work and raw material. Mills to reduce the palmetto, bear grass, and other fibrous materials, to pulp for the manufacture of paper and a variety of other articles, could here find unfailing employ-ment. A mill for making oil from the castor bean could develop a profitable industry. Canning factories for the tomatoes, for making guava jelly, etc., will here find an extensive field. Orange wine will be the typical drink of the Floridian; manufactories are in demand. Cassava and arrow-root make excellent starch; their cultivation might be stimulated by manufactories. The people of every village need their wood sawed very short for cooking purposes; a portable engine and circular saw could have steady employment. Cement tile and artificial stone are in constant demand; they might be manufactured here and save the expense of shipment. Improved means of "grubbing" land, cutting ditches, etc., are in demand; here is a valuable field for the inventor. South Florida has no book bindery; one is much needed. Paper mills are in demand; plenty of the raw material grows wild. An ice factory is needed, to save the expense of shipment. Wood-working machines of all kinds can find steady employment in working up the great variety of timber. In brief, there is room and oppor-tunity here for nearly every known industrial occupation, and the field awaits men of pluck and enterprise.

RAILROADS.

Of these promoters of development and necessities of civilization two are already completed to Polk County—the South Florida from Sanford, on Lake Monroe, to Tampa, on the Gulf of Mexico, with a branch from Bartow Junction, seventeen miles, to Bartow; The Flor-ida Southern from Lake City to Lakeland, gives railroad connection with Jacksonville and the whole railroad system of the country. This is to be speedily extended through Polk County to Charlotte Harbor, the date of completion being fixed at January 1, 1886. Work is pro-gressing. The survey runs through Bartow and Fort Meade.

The Tavares, Apopka and Gulf Railroad, which connects with the Florida Railway and Navigation Company's system, has been commenced, and sufficient iron contracted for to lay the track to Fort Meade. This road is to be built to Charlotte Harbor, with branches to Kissimmee, the lakes southeast of Bartow, Manatee and Fort Myers, making it a grand trunk line, running north and south through Polk County, near the centre.

Roads chartered are the Tropical, or Peninsular ; the Jacksonville, Tampa and Key West, which are to run north and south through the county ; the Bartow and Tampa ; the Indian River and Manatee, from Titusville, Brevard County, via Bartow and Fort Meade, to the mouth of the Manatee River ; Fort Meade, Keystone and Walk-in-the-Water Railroad, with a branch from Keystone to Arbuckle River, and the Bartow and DeLeon Springs Railroad. In addition to those already chartered, several others are contemplated, and will be built, doubtless, with but little delay, as the railroad system of Polk County gives great promise of being very intricate and complete.

GROWTH OF TOWNS AND VILLAGES.

Briefly noting that Florida came into the possession of the United States in 1821, we will defer consideration of its early history to a future chapter. All South Florida, with the exception of a few seaports, and a considerable portion of the northern part of the Territory, was in the virtual possession of the Indians until the breaking out of the Indian war in 1835. The war ended in 1842, military posts having been established about twenty miles apart throughout a large portion of the peninsula. In 1845 Florida was admitted as a State into the Union. In 1852 Fort Meade was occupied by a garrison of United States troops, but the whole country was a wilderness. About this time settlers, and especially cattlemen, began to settle on the fertile lands and pasture their cattle on the luxuriant ranges. It was then a part of Hillsboro' county.

In 1855 another Indian war broke out, but was ended in 1858 by the emigration of most of the Indians to beyond the Mississippi, the General Government paying $250 in gold for each warrior, and a less amount for the squaws and pappooses.

In 1859 Polk County was formed by a division of Hillsboro' County.

In 1861 the Florida Legislature passed an Act of Secession, an d cast her lot with the Southern Confederacy, and even the slow development of the county virtually ceased.

In 1865 the Act of Secession was repealed, the war having ended, and a new Constitution was framed and adopted by the State; but time was required to recuperate from the effects of the war before there could be substantial progress.

Efforts had been made since the organization of the county, in 1859, but without success, until Jacob Summerlin, in 1866, donated forty acres of land—the present site of the business portion of Bartow—to the county, for school purposes and a county site. He also gave twenty acres each to the Baptist and Methodist religious organizations at the same place, then known as Pease Creek. A courthouse, hotel, stores and several other buildings were built that year.

Time passes on; the population slowly increase, a few orange trees are set about the scattered residences; corn, pease, sugar-cane, cotton and a few vegetables are raised, but the chief wealth of the county is in the numerous herds of cattle that feed upon the luxuriant ranges of Polk, Manatee and Monroe Counties, and are exchanged for Spanish gold, the cattle being shipped from Punta Rassa to Havana. Tampa, forty-five miles to the west, was the entrepot and chief centre of trade, but Bartow and Fort Meade each had a couple of stores that did a heavy business.

The northern and eastern part of the State, accessible from the St. Johns River, had made good headway in its development. The South Florida Railroad, the first in South Florida, had been built to Orlando in 1880, and opened to Kissimmee in 1882, and new settlers were pouring in by hundreds, and prospectors by thousands, but Polk County was shut out from the activities of other parts of the country because of her lack of means and ways of transportation. There was a rough, unabridged wagon road forty-five miles to Tampa, and a trail through the woods, seventy miles to Orlando. There was no sale for fruit or for farm produce, except corn, because of the difficulty and expense of getting them to market. There were few immigrants because of this same lack of transportation facilities.

The more clear-headed and energetic of the inhabitants, feeling assured that the sterling virtues of the climate and soil of Polk County would eventually be made accessible, and be in great demand, very wisely went to planting groves. Though shut out from the busy, bustling outside world, the people were self-sustaining, happy and contented Talk of railroads was rife, and the survey of the South Florida Railroad was made from Kissimmee to Tampa, and its construction commenced. Prospective settlers and the agents of capitalists swarmed over Polk County, and many thousands of acres of land were purchased of the General Government and of the State. Hamilton Disston, who had purchased four million acres of the State, located large tracts in this county, while thousands of acres were reserved for chartered railroads.

In 1882, land was held at very low prices at Bartow and throughout the county. In 1883, it began to advance. The surveyed railroad was building through Polk County, about fourteen miles north of Bartow. In 1884, it was open to the public. A branch road was surveyed to Bartow. The real building of the town had hardly commenced in 1883. In 1884 it was earnestly prosecuted. Things began to boom. January, 1885, the branch road was opened by an excursion. The people of Polk County welcomed the guests with a magnificent barbecue. So abundant was the repast that at least seven times seven baskets of fragments must have remained. There was music by the Bartow brass band, a procession, speeches of welcome, music, a rare feast, excellent horseback riding by ladies and gentlemen, and the day closed with a dance at the leading hotel and a performance at the new opera-house. The visitors were agreeably surprised by the numbers of new buildings and the many unexpected evidences of progress. Polk County had begun her development, and she commenced strong. During the year past, the pine woods have been felled, and several vigorous new towns have sprung into being

along the line of the railroad—Lakeland, Acton, Auburndale, Sanitaria, Bartow Junction, Haines City, Winter Haven. Bartow has increased prodiguously, and Fort Meade, always an important centre of
trade, is making ready for wonderful strides when the railroad, or railroads, reaches there the coming season. Other centres are also pre-
paring for rapid development.

Thus is stated, as clearly and as briefly as possible, the Polk
County of the past and of the present. Is any further explanation
needed of the fact that she is not as densely populated as the one with
which she has the most points of similarity—Orange County? Polk
had her first railroad last year; her branch to the county seat this year.
Orange has had her railroad to her county seat for five years. The
development of Polk County in the next five years, judging from
present indications, will greatly surpass anything that has been seen
in Florida. It has the soil, it has the climate, it has the variety of
configuration of land and landscape, it has the locations for homes, it
has the opportunities and advantages for self-support, for wealth and
for fortunes, that is without a peer in this broad land. It is unequalled.
She is not laggard in her progressive speed. She has started on the
race of development with the strength of a giant and the vigor and
agility of an athlete.

PROSPECTS OF THE FUTURE.

Can anything be said that will enhance or show more clearly the
solid, actual prospects of so healthful, so beautiful and so fertile, so
attractive and desirable a county as that of Polk—a section where it
is a delight to live, where life is easily sustained and where corroding
care has no place?

Can we picture the near future? A dense population, each family
occupying from one to five or ten acres for the home lot, which is covered with fruit trees, prominent among which is the beautiful ever-
green orange tree, laden with its luscious and wealth-producing golden
fruit. The houses are embowered in running vines and the yard is
filled with beautiful shrubs and entrancing flowers, that grow and
bloom the livelong year. Near the kitchen is a plat devoted to the
vegetables, which are supplied fresh to the table every day in the
year. Also, we see a strawberry bed laden with delicious fruit from
December to June. There are also a great variety of fruits, for these
healthful products of Nature's alchemy form a pleasant portion of the
daily sustenance. There are grapes and figs, plums and peaches, varied fruits and great store of berries. But we will not enumerate;
there is profusion and abundance—everything that may delight the
taste or satisfy the appetite. The house is neatly built and elegantly
furnished. Abundance of windows and doors, wide halls and broad
verandas, giving free access to the balmy air, enable the happy owner
to banish exhaustive care and enjoy life to the full in the most delicious and healthful climate that the world affords, where Nature
clothes the earth in the most entrancing garments of beauty.

The near future will see active social centres every two to four
miles, with post-office, telegraph, telephone, express, and other desirable offices; with church, school, stores, etc., and most likely a railway depot. Not only will trains be run on the intricate net-work of

steel roads, but the railroad tricycle, propelled by foot or by electricity, will give individuals and families opportunity to go where and when they will on the regular lines, or on roads built for the purpose. The chief industries will be the growing of numerous varieties of fruits and vegetables for export to less favored sections, the entertainment of the thousands of visitors that will flock here at all seasons of the year, and for whom the most elegant of accommodations and the greatest variety of means of enjoyment will be provided. Thousands of people will find profitable occupation in the great variety of industries necessary to supply the needs, wants and desires of the great masses of highly civilized people that will here make their homes. Educational institutions of a high order will be numerous, as the genial and healthful climate will be found more favorable to study than that of any other section. Hence, knowledge will increase.

ADVANTAGES POLK COUNTY OFFERS.

1. The most delicious climate in the known world, with exemption from destructive frosts and freezes.

2. The most healthful section, as well as a varied, beautiful and productive country.

3. Fertile lands especially suited to the growth of citrus and other semi-tropical, as well as many small fruits, grains, etc., and the whole range of vegetables, all of which yield abundant and profitable returns.

4. Opportunity to work every day in the year, if desired, and that, too, more comfortably than in any other section, the summers being cooler and the winters warmer than in other places.

5. No long, cold, stormy winters to exhaust the products of the summer's industry, but instead, bountiful harvests every month in the year.

6. The best possible opportunities to easily secure not only a livelihood, but also a competence or a fortune.

7. Choice lands that can be secured at very moderate prices, as compared with their prospective value in the near future.

8. Good society; social centres with schools, churches, halls, stores, etc., and a rapidly increasing immigration of the most intelligent, cultured, earnest, energetic, temperate and law-abiding class of citizens that the whole country affords.

9. Ready means of communication by telegraph and railroad with all parts of the world, and rapidly increasing transportation facilities throughout the length and breadth of the county.

10. Temperate, orderly, progressive, social society, rapidly increasing educational and religious facilities, low taxes and the best possible indications of a dense population and abundant wealth.

11. A country comparatively free from insect pests, poisonous and dangerous reptiles and other common sources of annoyance.

12. Polk County is a section where the diseases are few and mild, yielding readily to proper treatment; where the death-rate is very light, and old age is the most fatal affection; where typhoid and scarlet fevers, pneumonia and phthisis, prevalent in other parts of the world, are very rare; where diphtheria, yellow fever, hydrophobia

and sun-stroke are unknown; where health and wealth, joy and prosperity abound; where pleasures are many and real discomforts are but few; where labor, energy and intelligent enterprise secure abundant rewards; where new industries may be established with a certainty of profitable returns; where children thrive and grow strong and vigorous, untouched by croup, diphtheria, scarlet fever, and other virulent diseases, that whelm with grief and sorrow so many families in other parts of the country; in brief, the great attractions of Polk County are in the fact that it is a land of balmy breezes, genial sunshine and active health, of delicious joys and increasing wealth.

CLASSES OF PEOPLE WANTED.

People of every degree, the rich and the poor, the educated and the illiterate, even the rheumatically lame, and the constitutionally lazy are invited to make their homes in Polk County, provided, always, that they are neither rascals, loafers, nor parasites. All such are advised to go to hades or sheol at once without taking the trouble to come to Polk County, as they would find themselves on the wrong road.

The rich can here increase their fortunes; the poor can secure competence and independence; the rheumatic will be healed; also, the consumptive, if not in the later stages, and the constitutionally lazy will find that the climate infuses so much life and healthful energy that even he will become ashamed of himself and be inspired with an ambition and earnest desire to also have a productive grove and a beautiful home.

Literary people will here find a very congenial field, for here they can meet with the intelligent, the educated and the cultured from all sections. Here they will find many phases of character that they can study with profit. They will also here find great variety. Abiding pleasantly in some of the many lovely locations that Polk County affords, they can store their minds with facts from varied sources. For exercise they have their choice of charming walks, rowing, or sailing on the beautiful lakes; they can hunt or fish for occasional recreation, or they can attend the numerous picnics and thus divert their minds. Are they lovers of botany, ornithology, ichthiology, or any of the ologies of natural science, they will here find an extensive and wonderfully interesting field. Or, are they of a yet more practical turn of mind, they can find abundant needful exercise for the body in the care of a charming grove, or a yard of lovely flowers, or a garden of toothsome vegetables, though devoting the bulk of their time to their literary labor in this genial climate.

EXTENT OF FLORIDA.

Florida is so large a State, extending nearly 400 miles from north to south, and about the same distance from east to west, with an area of 59,268 square miles, making it the largest State east of the Mississippi River; with a territory within less than 100 square miles as large as the important States of New York, Connecticut, New Jersey and Delaware (a fact that is realized by very few), covering, as it does, about the same number of degrees of latitude and longitude as do those four States, that people are apt to get very wrong ideas, not only of the State as a whole, but, also, of special localities, some of the individual counties being as large, or larger, than some of the States of the Union. Hence the many absurd ideas that prevail throughout the country with regard to Florida. A report, good or bad, may be absolutely true with regard to a particular locality, yet utterly untrue as regards many other parts of this great State, portions of which are destined to become the sanitariums of the Union, and the State, as a whole, very populous and wealthy.

If a person were to enter New York at its northern boundary, and make his way mostly by boat as far south as Lake Champlain, without visiting the interior country, and write letters to the papers with the idea that he knew all there was to be known with regard to New York, Connecticut, New Jersey and Delaware, would you consider his statements entitled to belief? Yet his case would be no more absurd, ridiculous as it would be, than is that of hundreds of tourists, or prospectors, who, reaching Florida by rail or water, make a trip by boat up the St. Johns River and return, thinking themselves thoroughly posted with regard to Florida. The northern New York traveler would have a much better idea of the characteristics and resources of those four States than would the Florida traveler up the St. Johns, for the characteristics, resources and productions of the different parts of Florida are much more varied than are to be found from northern New York to southern Delaware.

Hence, the advice, "Come and see for yourself," is very appropriate as regards Florida. Extensive tracts in the other States have special characteristics of similarity. So has Florida, the special characteristic here being that things are inextricably mixed. One man here may possess a perfect Eden, while his neighbor on the adjoining quarter section, as compared with the good fortune of his neighbor, may be in hades or sheol, as the new translation makes it. The good and the bad, the bitter and the sweet, are in close juxtaposition in Florida, and it requires practical common sense, and some acquaintance with the country, to determine the more desirable location. There is much waste land, consequently really good and desirable locations are sure to always command a high price, and are certain to be good investments.

Judge J. G. Knapp, whose writings in the *Times-Union* show a very intimate knowledge of Florida, divides the State into eight distinct zones, the general characteristics of each zone being specifically different. The northwestern part of the State he classes by itself. Then comes, proceeding from north to south, the Northern, the North Central, the Central, the South Central, the Southern, the Semi-Tropical, and lastly, the Tropical Zone, embracing the outlying islands or keys. Each division embraces a degree of latitude, except the Northern and Southern, the State extending from 30° 40′ 18″ to 24° 33′ north. The longitude west from Greenwich is from 80° 10′ to 87° 18′ 23″. His division is a very convenient one, and as reliable as it is possible to indicate special characteristics by arbitrary lines. There is, in fact, a gradual change from north to south, his division indicating the prominent features.

THE SOUTH CENTRAL BELT.

Regarding this belt of country, of which Polk County consti-
tutes the northern part of the centre, Judge J. G. Knapp, a standard
authority, says: "If we were charmed above the twenty-eighth paral-
lel, in this South Central Belt we will be enchanted. The climate here
is still more salubrious, more equable, the breezes more balsamic and
refreshing, the vegetation more tropical, more luxuriant, more constantly growing, blooming and fruiting; a summer where the sun
does not burn by day, and hot winds do not blow by night. Here
the planting season begins January 1st, and ends December 31st.

"Here frosts never freeze the orange, and it may remain on
the parent tree till fully ripe. The grape-fruit mellows and ripens in
April sunshine, and sweetens in the warm days of May. The lemon
swells in size and fills with its luscious acid under the dews and bright
sun of winter. The limes grow under the warm breath of spring and
mature their medicinal juices in June, when fevers commence to rage
in colder climes. Here grow the melting sugar apples, the sops and
cheramozas, and mangoes, with the rich guavas, that ripen in summer
and autumn, and thus lengthen out the season of the bananas and
pine-apples. Towering over all of these is the 'alligator' pear, half
vegetable, half salad, strange compound fruit of a tropical laurel.
Roses bloom eternally, and all the region is one vast green house, not
yet filled with plants, because man has not been placed here to dress
and care for it. The way hither has, until lately, been blocked by
savages, by false reports of innumerable pests of flies, snakes, alligators and wild beasts, is now just broken by a single railroad. The
beacon of truth begins to shine brightly aloft. The Nation's eye is
turned thither, the tread of the pioneer is heard, and men are making
their way there to stay.

"In this belt the backbone of the peninsula sinks down towards
the flat-woods and grassy meadows of Southern Manatee and the Ever-
glades. The prevailing winds are easterly, and show that the tropics
are near. The dry lands are covered with the stately southern pines,
the best fruit lands of the State, only needing the hand of man to re-
move the trees and fill its sandy, well drained soil with vegetable
matter. Hammocks as rich as any in the State are stretched along
its rivers and their branches, and elsewhere. The flat-woods are by nature rich grazing grounds, and can be cheaply converted into ever-
green meadows and pastures. Sugar-cane matures to tasseling, and
only requires a fertile soil for its production. Rice on suitable soil
will yield more bushels to the acre than can be produced on the rich,
black lands of the Northwest of oats and barley. The sweet potatoes
and cassava roots ever lie in the soil of their growth. Irish potatoes
and garden truck are planted and grown when other regions are bound
in fetters of ice, and water-melons are ripe in April. He who shall
predict for this region a high rank among the incomparable belts of
Florida will not err. Its seaports are in Tampa Bay."

Being credited by the general public, who are conversant with
the facts, with having given clear, reliable and unexaggerated statements with regard to the localities I have previously described, it will
be my most earnest endeavor in the present work to maintain

the same high and trustworthy reputation, which I value more highly than aught else the world can bestow. My aim is to inform and benefit, not to mislead. I believe that there are many thousands at the North and in the great West who, did they know the exact facts, would make their homes in Polk County at the earliest possible date. Thousands, and tens of tens of thousands, who are engaged in a ceaseless struggle to keep the wolf from the door, or to lay by something for a " rainy day" can hardly be made to realize how greatly they could better their condition, their prospects and their fortunes, by coming to this genial clime and securing a few acres of the fertile soil of Polk County whereon to make a home, cultivate and improve the same.

Extended observation convinces me that the first step toward success in Florida is to be secured by the purchase and improvement of land. There are several reasons for this fact. If a man owns the place where he resides, he will busy himself many odd hours, that otherwise might not be utilized, in making small and desirable, though not necessarily positively-required improvements. All the members of his family, if it be a well regulated family, will be inspired with a commendable zeal to make the home as beautiful, comfortable and attractive as possible. The home place will be a savings bank, wherein are treasured the spare moments and hours, the spare dimes and dollars. The more perfect the place becomes, the more earnest will be the endeavor.

Secondly, very few have any just idea of how valuable a place of from two to five acres can be made, nor of how great an income can be secured from such a tract by proper cultivation and attention. The intensive system of cultivation is the only one that is profitable. The largest crop that can be grown on one acre pays much better than an ordinary crop from a number of acres.

The amount of fertilizer that can be derived from the slops and refuse of a house, the privies and the droppings of the poultry, if properly composted, is sufficient, if applied to Polk County soil, to grow all the eatables required for the family, and leave a surplus for sale to purchase other desired articles.

Another important fact is that in Polk County you can raise the citrus and other semi-tropical fruits that yield many times as much income per acre as do the farming lands of the North and West. An income of $500 to $1,000, and even double those amounts, per acre, is no uncommon thing here, whereas, in other parts of the country, the standard crops do not average a tenth of the amount, though requiring much more labor and care. Such encouraging results are attained by the growing of vegetables, as well as by the production of pineapples, oranges, etc.

The superior healthfulness of Polk County should be an important inducement to the settler. The vitalizing sunshine, the balmy and delicious atmosphere, the genial days, the restful nights, the absence of acclimatizing diseases, of phthisis and of zymotic diseases, as also of those maladies that are so fatal to children in other States, show very significantly the superior desirability of this section in preference to localities in other States of the Union.

A family can live much more cheaply, as well as comfortably, in Polk County than in other parts of the country, and can earn more.

There are no long and inclement winters to consume the substance of the harvest. There is no season of enforced idleness, but a man can work every day in the year if he choose. There is not a month, a week, or a day, but some of the products of the soil may be harvested, and he may, if he desire, have pleasantly varied kinds of vegetables fresh from his garden every day in the year. Here he is neither pinched with bitter cold nor exhausted by sultry days or nights, nor is he consumed by exhaustive heat. Modified by genial breezes, the temperature is refreshingly delicious at all seasons of the year. Not only are the winters mild, bright, healthful and inspiriting to active duties, but the summer heats, though extending over a longer period, are less intense and less exhaustive than in any other State. Besides this, however hot the day, the air never becomes close, sultry and putrid, as is so often the case in the interior States of the Union; however hot the day, the nights are invariably cool, refreshing, and conducive to healthful, strengthening sleep. This is due to the peninsular location of the State, and also to the fact that the days and nights are of more equal duration, giving less time for the earth to become heated by the sun's rays, while the nights being longer, it has more time to cool. The peninsular form gives semi-daily breezes, the trade winds from the Atlantic by day and the reverse winds from the Gulf by night. But the trade winds are only felt in the peninsular part of Florida. Polk County being located near the centre of the peninsula, it gets the full benefit of the trade winds, which are a very important factor in securing for it the unexcelled deliciousness of its climate, that so favorably impresses all who have the good fortune to make its acquaintance. In this respect Polk County claims the pre-eminence over all other sections.

Provisions and general merchandise are supplied as cheaply and as reasonably in Polk County as in most sections of the Union. This fact is a surprise to many who come here. Supplies are received both from Atlantic and Gulf ports, as well as from the interior of the country. A man who can get a comfortable living at the North or West can secure one here with half the effort. He has every day in the year to work, the fertile soil yields abundant returns, the semi-tropic fruits and vegetables give a much greater cash value per acre than the Northern and Western farm crops, fuel costs only the cutting and hauling, and but little is required, less clothing is needed, doctors' bills vanish, taxes are light, encouraging prospects drive away the blues, and there seems to be no possible reason why every sensible and intelligent person should not only enjoy life to the full, but also leave an increasing competence to his children after him.

The people who come to Florida are mostly people of some means, those in moderately comfortable circumstances, who desire not only to better their fortunes, but also to escape the rigors of the inclement Northern winters. The larger proportion are past middle age and have families. They are chiefly from the better classes of the industrious people, who are not afraid to work—earnest, intelligent, and many of them highly educated.

There are specimens only of the other classes that compose society. Those who come to Florida to make their homes are, as a class, greatly superior to the immigrants to the new Western States, socially, morally, and intellectually. No other State was ever able to secure a class of people like those who are pouring into Florida by thousands. They bring their home ideas with them and put them in active operation at once. They are energetic, industrious, orderly, God-fearing people, the very cream or selection of the fittest from the same classes at the North and West, the Middle and Central States, and the South. With a steady and rapid infusion of such fresh blood poured into Florida by many streams, and in augmenting quantities, is it any wonder that her fast-increasing population begin to feel within their veins the pulsations of the grand and sublime destiny that awaits Florida in the near future, and that will place upon her imperial brow not only the victor's crowns of laurel, of oak, and of bay, but also the most sparkling, the brightest, and the richest diadem that the genius of the age can produce? Nature and art, wealth and culture will here unitedly labor to create the grandest achievements possible to the human race.

The people of the whole country, of the civilized world, in fact, are attracted to Florida as a haven of healthful rest, and yet of busy, enterprising activity. Few, indeed, are the localities on the whole wide earth's broad domain where health and wealth can be simultaneously pursued and secured, and that, too, in a land where the surroundings are the most satisfying possible; where every sense, feeling, aspiration and emotion that can give pleasure, or solid profit, is brought into healthfully active exercise. A land like Polk County, where severe and destructive diseases are unknown; a land where the soil is as fertile as it can possibly be, consistently with universal health; a land where the soil responds with royal munificence to the efforts of the cultivator, and where the variety of products, of grain, of vegetables, and of fruits is astonishingly large; a land of fertile fields, beautiful lakes, charming valleys and running streams; a land abundant in fish and game; "a land flowing with milk and honey;" a land where the water is pure, soft, healthful, and easily obtained; a land whose soil is firm and compact, surprisingly free from dust, and mud, and stones; a land of thrifty timber—beautiful pines and sturdy oaks predominating among the scores and scores and scores of other native trees; a land of running and fruitful vines; a land abounding in berries, in grapes, in a variety of wild fruits; a land where the fruits and vegetables of the semi-tropics, as well as of the temperate zones, thrive, grow vigorously and produce abundantly; a land where the mornings are unspeakably fragrant and delicious, the days radiant with bright sunshine, pleasantly obscured at intervals with beautiful clouds, that flit like watchful and beneficent angels athwart the face of the heavens, while gentle breezes cool and renew the air. The evenings have a charm peculiarly their own—soft, soothing, restful, promotive of a love-like, delicious languor that well prepares one for a restful and refreshing night's sleep, after the labors and pleasures of the day. Of the nights themselves, it is impossible to overstate their charms, their solid, substantial advantages, as compared with the nights in other sections. Until realized by actual personal experience, no one can form any adequate conception of the healthful luxury that night gives. "He giveth His beloved sleep." Here the nights subserve their true purpose. Sleep comes as ordained by nature—restful, refreshing, strengthening. The sleeper awakes with the early dawn, thoroughly renewed in body and in mind, fully prepared for the duties and the privileges of the coming day.

In Polk County there is no wearied tossing on burning couches, vainly seeking rest, gasping for a. breath of fresh air in a sultry, oppressive, stifling atmosphere, as is so often the case in interior countries, during the heated term. On the contrary, as soon as the sun sets, a delicious sense of coolness pervades the atmosphere, permeating every locality to which the balsamic perfume-laden air has free access, increasing in degree as the night advances, until it culminates in the glorious deliciousness of the early morning, calm, tranquil, freshly odorous, and inspiring worthy thoughts and noble deeds.

DIVERSIFIED APPEARANCE.

The face of the country is varied, though neither generally flat nor hilly. The predominant growth is pine, though often interspersed with oak of different varieties, persimmon, wild cherry, etc., which gives a very pleasing effect. Along the water courses, sweet and black gum, maple, cypress, cabbage palm, wild orange, whitewood, etc.. are the chief growth. The numerous lovely lakes are a great attraction. The soil varies in different localities, but a large portion is of excellent quality, producing remunerative crops, without the aid of commercial fertilizers. Ever since the first settlement, some thirty years since, the people have secured a generous support from the raising of cattle and cultivation of general farm crops.

SOIL, WATER, CLIMATE.

The land is peculiarly adapted to the growth of citrus fruits, and their quality is of the very finest. Strawberries are also a great success, as well as garden vegetables of all kinds. Fibrous plants also grow vigorously. The soil is a sandy loam, underlaid at varying depths with a stratum of clay. Water of excellent quality is secured at differing depths, the well diggers passing through several strata of alternating sand and clay, until the water-bearing strata of fine gravel and sandy clay is reached. The low latitude insures short and mild winters, sometimes without frost,. while the winds from Ocean and Gulf temper the heats of summer, making them much less oppressive than farther north, the extremes of temperature never being so great, insuring deliciously cool and restful nights.

The soil of HOMELAND is the choicest that can be found in the most healthy portions of healthful South Florida, and responds readily to the cultivator's demands, producing a majority of the fruits, vegetables, flowers and woods grown by the civilized and enlightened nations of the whole world.

In HOMELAND can be brought to profitable maturity a greater variety of products than any other section can boast. Its soil is so varied that here can be found every kind and quality that may be desired, high or low, rich or poor, all adapted to special and varied products. Here are rich and productive fields, thrifty pines, vigorous and enduring oaks, and a hundred or more other choice and desirable forest woods.

In HOMELAND are an abundance of lovely lakes, running streams, beautiful valleys, grass-covered hills and plains, that respond readily to the efforts and ambitions of those who may seek thereon to improve health or fortune. It is not a fairy land, but it is a land of practical and substantial realities. The earlier and the later settlers here have been self-supporting from the first. Nor have they had to endure the hardships of the pioneers of other sections. The land has not been hard to clear, and it has responded readily to the cultivator's art, the earth producing abundantly.

But, remarks the reader, this is all very pleasant and encouraging, yet, I have often read glowing and enchanting accounts of other sections that made them appear a veritable paridise. The stern realities of bitter experience, however, dispelled these illusions and taught caution; how shall I be assured to a certainty that Polk County is all that it is pictured in these pages? No one section has all the good things. The bitter ever mingles with the sweet, the evil with the good. It is thus throughout the whole realm of nature. What we denizens of other localites desire are the exact facts, the bad as well as the good, the unpleasant as well as the attractive features.

The writer of these pages knows from his own experience that demands like the foregoing will be made, and justly, therefore he will earnestly endeavor fairly, squarely and conscientiously to present as plain, truthful and unexagerated a picture of Polk County as it is possible to make.

Before coming to Florida, in 1881, he had had an extended and active experience for nearly half a century in the Eastern, Middle and Western States, as well as in Delaware, Maryland and Virginia, and therefore is not only well acquainted with the characteristics, resources,capabilities and advatages possessed by those important sections of the country, but with their disadvantages also. He well knows the character, aims and aspirations of the Northern and Western people, the difficulties with which they have to contend their annoyances, the hardships consequent upon the storms, drouths and vigorous inclemencies of the seasons, and compare them faithfully and honestly with the conditions in this and other parts of Florida. His locks have been duly bleached by the frosts of many winters, and his enthusiasm tempered by experience and adversity. He may not have lost his youthful fire and earnestness, but he has been schooled by the cares and duties of active life to close observation, theughtful reflection and carefulness of expression.

For several years before coming to Florida he received the newspapers of all sections of the State, as exchanges, and studied them carefully ; also, many books and pamphlets regarding Florida. His health steadily improving since reaching here, he has personally studied South Florida from the Atlantic Ocean to the Gulf of Mexico, the results of his observations being published in various Northern newspapers and magazines, and in the local press of South Florida. He was also the author of "Orangeland," the Orange County pamphlet, and has devoted his time and abilities to securing as thorough a knowledge as possible of the characteristics, resources and advantages possessed by the different sections of South Florida.

Thus far, no one has accused him of exaggeration or misrepresentation, and his constant aim has been to neither flatter the country nor the people, but to give such unequivocal information as may be of the greatest benefit to inquiring minds wherever they may be located. In writing this work, at the solicitation of leading citizens of Polk County, who desire the exact facts to be made known, the same rule is observed.

" But," the interested reader will say, " you have given the poetry of Polk County; give us the plain prose. We shall be glad to enjoy every delight, every comfort and luxury that Polk County has to give, but the first practical question is, how shall we so invest our moderate means that, with our labor, we can secure a livelihood and an increasing competence for our old age and for our children."

The question is a pertinent one to the majority who come to Florida, to Polk County, to the true HOMELAND. The wealthy—and many of this class are coming to Florida—can get along without difficulty anywhere, yet I have noticed that they are quite as careful in regard to their investments, and in the choice of a location for a home as are those possessed of less means. To them Polk County offers attractive inducements in the most delightful climate anywhere to be found, in the very choicest of building sites, with or without lake fronts, in an enterprising and very rapidly increasing population, nowhere excelled for good qualities, in ready access to the other parts of the country by the lines of enterprising railroads, in chances for profitable business, in sure and speedy increase of fortune by invest -ments in land.

To people of moderate means, the first necessity is a place to live. Unless they have acquaintances in other sections near whom they intend to locate, their objective point will be rapidly-growing Bartow the county seat. This is reached by the South Florida Railroad, either from Tampa, on the Gulf of Mexico, or by way of Jacksonville, from Sanford, on Lake Monroe, at the head of navigation, by large steamboats, of the St. Johns River. Or, the Florida Southern Railroad being completed from Jacksonville to Lakeland, in Polk County, where it forms a junction with the South Florida Railroad, will be a favorite route with many, as it runs north and south through the interior of the State. It is expected that this road will be extended to Bartow, Fort Meade and beyond, the present season, opening up one of the very finest tracts of country in Florida. The South Florida is soon to be extended also, and several other lines are chartered, traversing various parts of the county, and will soon be built. There is also an all-rail route from Jacksonville to Orlando, thence to Bartow—broad gauge as far as Orlando. This runs through the finest and most developed portions of the State.

Reaching Polk County, a few days will be judiciously devoted to examining the merits of different sections, and the advantages offered, by the varied properties. Bartow has good hotels and boarding-houses whose charges are moderate—one to two dollars per day, but less by the week—and there are active real estate agents who will gladly show you the country.

Everything looks strange to the new-comer. The contour of the country and the vegetation is different. The soil, a sandy loam, underlaid with clay in the best sections, is totally unlike the soils of the

North and the West, yet the growing crops and the thrifty and abundant vegetation prove that it has surprising elements of fertility. There is a firm, substantial tread to the land, totally unlike that of many other sections of Florida, which is a pleasant surprise to the visitor. Though essentially a pine country, the abundance of vigorous oaks that greet the eye in every direction, on the uplands as well as in the valleys, give a character of solid, substantial permanence that confirms the impression that this is the best and the most desirable portion of Florida, and that it will speedily be thickly dotted with beautiful homes. It is the true HOMELAND, soil and climate, productiveness and healthfulness, with comparative freedom from troublesome and annoying insect pests and harmful reptiles, combining to make life here not only endurable, but delightful, profitable, and pleasant. A livelihood is easily secured here, while a competence or a fortune requires but little effort, when combined with active and intelligent judgment. As the prospector goes over the country he finds a great variety of locations. To secure the most profitable requires not only intelligent judgment, but assistance from those acquainted with the country.

Where to locate depends upon the designs, the contemplated business of the locator, his desires as regards society, and the amount of capital at his disposal. If he designs engaging in mercantile business, or kindred pursuits, he should locate either in a town, or where a trade centre will soon be established. Bartow is the county seat, and is likely to remain such as long as Polk County exists. Therefore, there is a fixed centre that is sure to increase rapidly in population, while landed property will constantly augment in value. It has also a superior back country surrounding it, that will make this the grand centre, shipping and receiving their supplies from here, and here establishing new industries.

There is no present prospect that Polk County will have any large cities, as compared with cities at the North, but there is every indication that it will soon be very thickly populated, with a residence on every acre or two of the more healthful and desirable lands, the produce from which, with the intensive system of culture, will be sufficient to maintain each family in comfort, and even in luxury. This will compel trade centres every three or four miles at the farthest, with post-office, telegraph, telephone, and express offices, schools, churches, stores, and public libraries. In many of these social centres varied manufactories will be established, to supply the wants of the community and provide articles for shipment.

People will come here in multitudes, as soon as the superior attractions of the climate, the agreeable healthfulness and the encouraging ease of securing a livelihood here become known, as they are sure to be in a brief period. There is great activity here now, but it is not a drop in the bucket, compared with what will soon be seen. Prices of lands, now moderate, will double and double with so astonishing rapidity as to be beyond the reach of all, except those of considerable means. Thousands will secure fortunes in the advance in the price of lands. Hence, those who are on the ground first will have the best opportunity to secure fortunes, with no effort except that arising from intelligent and judicious investments.

Polk County already has two railroad lines, the South Florida and the Florida Southern. They strike the county in its northern portion, and will extend through the county from north to south. These, with their branches, and with other contemplated lines, will bring every section within easy reach of railroad transportation to any part of the Union.

Many will deem it most desirable to locate along the lines already built. That will give a present convenience and a certainty as to the future, but the lands are held at higher prices. Along the route of contemplated railroad lines are many excellent bargains in land, the prices of which will greatly increase when the roads are built, which will doubtless be done at a very early period. Away from the rail-road lines completed, and those prospective, that are certain to be built, the very best judgment of the locator is necessary, in order to make the best possible investment; for, the qualities of the location being otherwise equal, lands within a mile of a railroad depot are sure to be of considerable more value than those farther away, though, at present prices, I know of no lands but will eventually prove profitable investments, and before the close of the present century it is probable that every desirable locality here will be within two or three miles of a railroad, at farthest.

Though Polk County has not had the advantages of railroad transportation to its county seat, Bartow, until the present year, and to no part of the county until the year previous; people who visit it are surprised at the extent and the rapidity of its development. There is here a character of progress and substantiability that does not fail to favorably impress every one who investigates the condition and the prospects of this section. The buildings of the earlier settlers were mostly small and unpretentious, as well as widely scattered. They were all that the few and simple wants of stock-growers and general farmers required in so mild and genial a climate. Most of them set out a few orange trees for home supply.

Few groves were set in Polk County until since the commencement of the present decade, but the few trees about the cabins showed that both soil and climate were peculiarly adapted to their vigorous growth and prolific yield of the delicious golden fruit. With the progress of the South Florida Railroad from Sanford, and the fast increasing certainty of speedy railroad transportation, numbers of groves were set, especially on that wonderful tract of beautiful and fertile country that extends from north to south a few miles each side of Peace River, Bartow being near the northern, and Fort Meade near the southern portion. Numbers of these groves are now yielding abundant returns, and well illustrate the great capabilities of the country, especially the few that were set eight or ten years since. To see is to believe.

The past two years have shown an immense and remarkable development in every respect. People are locating here from every State in the Union, and the population is doubling rapidly. Hundreds of new groves are being set, and many acres of land being rapidly cleared. New substantial, commodious and elegant buildings are being erected, that will compare favorably with those of any section of the State. In fact, no part of the country shows more active, rapid and substantial development. Its destiny is secure and glorious.

To decide upon the location where you will make your home is not the easiest matter in the world, as there seem to be so many things to be considered; so many conflicting claims, advantages and disadvantages. There is the nearness to, or distance from, a present or future centre, the comparative fertility of the soil, the relative healthfulness, the natural beauty of one location as compared with another, the actual market value of the land positive and comparative, the chances of the future importance of the several localities, the prospective main routes of travel, etc. All these things are to be considered, if one would do the best possible.

So many new and divergent ideas are likely to be crowded on the prospector's consideration that he will be fortunate if he is not attacked by mental dyspepsia or indigestion, and rendered unable to properly digest the heterogenous information received from all sources. There is one consolation and encouragement, however, and that is, if he does not locate in a low and undrainable locality, or on barren scrub or sand-hills, he is sure, at present prices, to make a good investment and one that will pay. Some will do even better than this, and secure prizes, but though all can be thoroughly assured of doing well, who will secure the larger prizes none can tell. Luck and chance, as well as judgment, here come into play. The unexpected location of a road, of a manufactory, or of some unanticipated enterprise, are things that the wisest cannot foresee, and this kind of lightning is liable to strike anywhere. Perhaps the new comer himself may be the conductor that will bring it to his location. To a certain extent, as in Oriental countries, "it is the unexpected that happens," mainly, however, the conditions are determined by sound judgment and active energy. To a great extent the new comer must rely upon the opinions of his intelligent and conscientious acquaintances who have secured a knowledge of the country by experience, yet it will not do to place too implicit confidence in any one. Not that they would intentionally misrepresent or mislead, but from the fact that every one soon gets very positive opinions here, and I have yet to find the first land-owner who did not veritably believe that, taking all things into consideration, he had the very best location anywhere to be found in the State. Hence, all who have been here a few years are happy, contented and hopeful. Most men are ready to sell a part of their holding that they may secure neighbors and the means for more extensive improvements; or the whole, because they know where they can reinvest their money to greater advantage. None sell to leave the country. They appreciate the country and its solid, substantial future prospects too highly to return to their former homes. They are here to stay and grow up with the country.

Having secured a location, the first thing to do is to clear, fence and build. Polk County is a land of substantial fences. Unless you have abundant means, better clear and fence only an acre at first. Do not go to work with the idea that you can do everything in a day, a month, or even a year. Make haste slowly. Build comfortably, but not extravagantly. Do not use up all your means in building, even though some of your neighbors may have a finer house than you can now comfortably afford. It will come by and by. Your money, if your means are somewhat limited, can be used to better advantage in making a grove, a pine-apple orchard, a strawberry field, or iu getting some other fruit well on the way to bearing. A little spare cash always comes handy. While getting started, your outgo will probably exceed your income. I am not now speaking of those extra smart people who, by sharp trades, can pay their way as they go, nor of those exceptionally fortunate ones who secure big crops the first year, and sell them at the very highest prices. Numbers do have such expeperiences here, but my remarks are intended for ordinary mortals, like myself, who secure results by patient and continued industry. All steady, persistent workers are sure to thrive, if they do not yield too much to their ambition, and overwork at first. I have had the experience, consequently I am entitled to give advice. You have every day in the year to labor, therefore don't rush to show how much smarter you are than your neighbors. People don't brag on big days' work, but the contrary, in this country. They could if they chose, but there is no use in it. There is plenty of time. If you have a few acres of land, the increase in value of that alone will make you rich. Smaller tracts, with groves properly cared for, will accomplish the same result.

The man who would succed here, and be happy and contented, should be a land-owner. He should be his own master. A man may secure a livelihood by working for others for wages, but no man ever got rich that way. It is frequently necessary to get a start by working for wages. Work and save and invest. Our most successful men have done that at first. You can do likewise if you are short of funds, but, if you have sufficient means to live, it will pay you much better to work upon your own place and improve it. If you are skillful and intelligent you will accomplish more by steady work and careful supervision than any man you can hire.

I do not mean by this that you should individually do all the necessary preliminary or later work, but that it should be done under your own careful supervision, at least. You should be employer, instead of employed, if possible. Some people, however, are only capable of working successfully by working for others. Others have a special talent or capacity in certain fields of labor that enables them to secure large wages. In such cases, I repeat, secure a tract of land and have it improved, even if you have to hire others to do the work while you are otherwise engaged. You have made a start in the right direction.

The main future dependence will be in your grove. " What," you ask, "can I do for a living while the grove is coming into bearing?" This is a pertinent and important question, for, with the best trees you can set for a grove—three-year-old trees—it will be five years before you can receive much income from them, and it will be several years more before they are in full bearing. It would be poor policy to sit down and wait all these years, simply cultivating the trees. But there are other matters that can profitably engage your attention, making the tree cultivation a mere incident.

Those who have tried it, have met with great success raising vegetables on the same land. In fact, the growth of the trees is promoted by the cultivation given to the vegetables. The only requisite is, that no more shall be taken from the soil than is returned in the way of added fertilizers. The best lands will bring a grove to bearing without the aid of fertilizers, but there is no land that is not benefitted by their application. On the best lands, however, but little is required. You could probably get along without any for several years, but its application would be judicious, and would pay well as an investment. Small fruits might be raised on a portion of your land. They would give quicker returns and would tide over the period of waiting for the grove to bear. Pine-apples will make returns the second year, and will find a ready home market at excellent prices. Strawberries set in September will give ample and profitable returns the following spring. They are peculiarly adapted to the soil and the climate. They not only find a ready market at home, but can be shipped North at a time to receive the very highest prices. But why enumerate? A variety of articles can be raised that will give quick returns.

Another important industry, the proceeds of which find a ready market, and give quick and profitable returns, is the raising of eggs and poultry. There is a lively and continuous demand for both, and they are never a drug in the market, but command ready sales at good prices. The climate is peculiarly adapted to this industry. Chickens can be hatched any and every month in the year, and thrive wonderfully with very little care. You must, however, either fence in your garden or your poultry. They are a great benefit to orange trees, as they scratch just about as deep as an orange tree ought to be cultivated. Their droppings are also quite valuable as a fertilizer.

Bees do extremely well, and those who have a taste for apiculture could secure quite a revenue from this source. Florida honey is not only equal to, but superior to, that of any other section. In this regard, even California has to take second place.

The raising of roses and other flowers for shipment to Northern cities is soon to be a very important industry. This pleasant occupation will give abundant and profitable returns.

Vegetables of all kinds grow rapidly and abundantly, and there is a rapidly increasing home as well as a Northern market. The leading standard articles for shipment are: cabbages, beans, cucumbers, tomatoes, beets, radishes, cauliflower, egg-plant, etc., etc.

Sweet corn, of the very finest quality, matures early, and might be placed in the markets of New York and other cities weeks before they could get a supply from any other source. The same is true of melons, squashes, etc. In fact, the lands of Polk County are peculiarly adapted to the speedy and early growth of all kinds of vegetables, and the enterprising can secure fortunes from that source alone, and another fortune from their orange grove.

What, then, does the fertile land and de'icious climate of Polk County offer to its active citizens as a source of livelihood, while growing an orange grove, that shall be a lasting fortune to them and to their descendants for many generations—offering advantages superior to any other part of Florida? Let us see, commencing with those industries that yield the quickest returns:

First.—There is the whole range and variety of vegetables.
Second.—Poultry are easily reared and give excellent profits.
Third.—Strawberries and other small fruits.

Fourth.—Bees and honey.

Fifth.—Roses and other flowering plants.

Sixth.—Pine-apples. Some winters these will need a slight protection a few nights against a light frost, as there is no such thing as a frost line in Florida. The low latitude of Polk County, however, gives almost complete immunity, and it can be readily seen that at the same elevation, the lower the latitude the greater the exemption. Its lake protection is also the most complete of any locality in Florida, being located chiefly in the northern and northeastern part of the county. It must also be remembered that these lakes receive the winds after much of the frost has been extracted by passing over the lakes in the counties still further north. Hence, if those sections have any exemption because of their bodies of water—and that they have is unquestionable—Polk County is, and must be, doubly protected. This is an important fact to be considered, and is one reason why the people of Polk have green corn, melons, strawberries, etc., etc., from Christmas to the 4th of July, as well as other fruits and vegetables at all seasons of the year.

Indisputable facts like the foregoing, once presented and duly considered, cannot justly fail to convince all, desiring to better their condition, that Polk County presents unequalled advantages as the inducement for them to locate within her borders.

Until the near advent of railroad transportation, the residents of Polk County, being chiefly stock raisers, had little or no desire for an increase of immigration, as all accessions of new settlers decreased the range for stock, and, having no facilities of transportation, no outlet by railroad or steamboat, they could derive no income from either fruit or vegetables, consequently they were disinclined to have the superior resources and advantages of this, the very choicest portion of the State, known except to their immediate friends. The South Florida Railroad has come, other roads are coming speedily, and the era of progress has commenced with greater vigor than in any other section of the State. The flow of the tide that leads on to fortune is rapidly increasing, and those who desire to secure the best possible advantages should not delay in their coming. A glorious opportunity is offered to all who desire to improve either health or fortune. This is pre-eminently the HOMELAND, and it will be speedily occupied by increasing thousands and tens of thousands. Here will be made the choicest, the most delightful homes that the whole world affords. Here will be found every luxury, every appliance of comfort and delight that the civilized world affords. Here is a climate without a peer—unequalled in winter and in summer. The winters are mild, delicious and delightful. The summers are cooler and more enjoyable than in more northern portions of the State. It is a peculiarity of this portion of the peninsula that should be considered and remembered, viz : That this section is warmer in winter and cooler in summer than the sections further north. No tornadoes nor cyclones have ever passed through this section, and its location is such that those devastating storms are here impossible. Being in the interior of the peninsula it is protected from the storms that rage on the Atlantic and Gulf Coasts. In fact, a more favorable location does not seem possible.

" But," says the reader, "you have given the bright and the pleasant, what are the darker and more sombre colors. Those are necessary
to complete the picture."

Let us analyze the matter carefully and plainly state every draw-
back, every disagreeable fact that prevents Polk County from being
a perfect Eden on earth. The unpleasant is not as agreeable to dwell
upon as the pleasant, but it is essential to the purpose of this work,
which is to represent things exactly as they are. Some of the days
are stormy, disagreeable, unpleasant, though the number of such days
is much less than in the continental parts of America.

A light fire is needed, a few, a very few, days in winter, when the
mercury drops below 60° or 65°, especially such mornings and evenings, to dispel the chill.

The rays of the sun during the mid-hours of the summer days
are very hot and piercing. The slightest shade interposed, however—
an umbrella, a tree or a roof, or a passing cloud—gives a refreshing
sense of coolness. In fact, however hot in the sun, the constant cool
breeze always insures a comfortably-cool place. The air is neither
sultry nor suffocating. The most oppressive time is usually from seven
to nine o'clock in the morning, before the refreshing effects of the sea
breeze is felt.

The frequent, almost daily, showers during the wet season of the
summer months are frequently inconvenient, but the occasional wet-
tings do no actual harm, not causing colds, as at the North, though it
is disagreeable to be wet. These showers cool the air and cause all
kinds of vegetation to grow very rapidly. They are actually one of
the good, instead of the bad features of the climate. The necessary
out-door work is the lightest during the hot and rainy season, instead
of the heaviest, as in other portions of the Union. The showers usually come in the afternoons of the hottest days, but there are no barns
to be filled with hay, and very little cultivation or harvesting is required during the summer months. Few crops mature during the
fervid heats of summer. No work is then pressing the one who is
fairly settled. It is a time of natural, not enforced, idleness, and is
much more agreeable than the necessary cessation of out-door
farm labor at the North during the cold and severe winters. Extremes
are disagreeable anywhere. Here the cold of winter is omitted, while
the summers are much more pleasant and enjoyable than at the North.
The same fact is also true of every other season of the year. Hence,
as regards the seasons, separately or collectively, Polk County has
greatly the advantage as regards Northern localities.

The labor of clearing land, felling trees, grubbing out roots and
making fences is no more difficult or disagreeable than at the North.
In fact, it is less so, as there are no stones in the way. The labor of
cultivation is also less, and much easier here. There is also no inter-
ruption from frost, ice and snow, which makes a great saving of time
and expense in favor of Polk County.

Here one loses the pleasure of doing the hardest and most exhaustive work during the hottest months, to fill barns, granaries and
cellars with hay, grain and roots to be fed to stock on the most frosty
mornings and the coldest and most disagreeable days.

The residents of Polk County lose the pleasure and profit derived

from frozen fingers, ears and toes. Another disagreeable matter is
the fact that ice has to be made artificially here. As nature is not
in the ice-making business the boys and girls cannot go skating on the
ponds, lakes and rivers, as at the North, but go boating instead at all
seasons of the year. Other luxuries that the youth here have to forego are those of sliding down hill, building snow-men and houses,
throwing snow-balls and fishing through the ice. Nor do they have
the pleasure of wallowing through snow-drifts, shoveling roads through
the snow or going sleigh-riding. They manage to do their courting
all the same, however.

Another disagreeable feature of Polk County, much missed by
Northern and Western people, is the speedy drying of the ground and
vegetation within a few minutes after each shower, which prevents
those disposed to take a walk, from enjoying the pleasure of wet feet
and attendant colds and rhuematism. Another thing much missed
is the lack of mud here, the porous soil absorbing the water instead of
permitting it to stand in puddles in the road. There is also such an
absence of dust that the course of carriages cannot be traced by
clouds of dust, and the occupants are prevented from enjoying the
familiar luxury of dust-baths so familiar to Northern and Western
people. There are other equally enjoyable luxuries common to other
sections that people here are compelled to forego, but those mentioned
will suffice.

> "Each matchless morning marches from the East
>   In tints inimitable and divine;
> Each perfect noon sustains the endless feast,
>   In which the wedded charms of life combine;
> Sweet evening waits till golden day, released,
>   Shall lead her blushing down the world's decline.

WHERE AND WHAT IS FLORIDA?

In deference to the wishes of the publishers, and for the convenience of those who have a thousand times read and re-read the history
of Florida, from the year of its discovery to the present time, I will
endeavor to give a brief outline of the more salient points, beginning
with the earliest records.

Geologically speaking, Florida is of very recent formation, although
that was doubtless millions of years ago. The learned savans say it
is like a book with only the title page and last chapter inclosed between the covers. That is to say, that its foundation, being the primitive granite, the usual succeeding formations are omitted until reaching the limestone and like strata of a recent epoch. As there has
been no thorough geological survey of the State, though one is much
needed and would unquestionably repay the cost a hundred-fold, the
statement may be considered as theoretical instead of an actual certainty. The State is, however, unquestionably one of recent development and still in process of formation. In process of time those
ndustrious workers, the coral insects, which, like the trout, love swift
running water, will undoubtedly connect Florida with the island of
Cuba; the Mississippi and other rivers, that year by year deposit
immense quantities of soil and drift in the Gulf of Mexico, will
eventually fill it, and Florida will not only cease to be a peninsula,
but will lose a great charm of its delicious climate—the cool night
breeze from the Gulf. This fact should be a warning to those
thoughtful ones who contemplate a permanent residence in Polk
County.

Advancing to pre-historic times we find that the Mound Builders
were the first inhabitants who have left any recognizable trace of their
occupation. Whether they were the first settlers, or whether then, as
in these recent years, an irresistible wave of immigration poured from
the North into this delectable land, must ever remain a matter of
conjecture. Let the wise men discuss the probabilities.

For aught we know, Florida may have been the site of the veritable Garden of Eden, some of the streams running in such a manner
as to make it easy for a skilful philologist and geographer to find
their counterpart in the description in Genesis of the streams from
whence "a mist went up to water the garden." In the land of
Havillah there was gold. That doubtless referred to Georgia. Noah's
ark doubtless floated across the sea to Mount Ararat. The length of
his voyage unquestionably gave sufficient time. I leave the full
elucidation of the matter to those deeply versed in Biblical lore,
without charge for the suggestion. We know that Polk County is in

about the same latitude as the supposed site of the sacred garden, and also, that it is in the same zone as the greatest and most prosprous cities of antiquity. This is another 'matter for profitable consideration.

Scientists tell us that the birth-place of the human race must have been near the sea-shore and in the southern edge of the North Temperate Zone. Both these conditions are better fulfilled by Florida than by any other portion of the known world, its system of interior lakes and water courses of both salt and fresh water, its numerous bays and lagoons, and its very superior climatic conditions, giving it the precedence over all other localities in the matter of desirability. These facts being duly considered, it becomes evident that here is a favorable locality for the genius of the nineteenth century to create and develop one of the most prosperous, wealthy and powerful of States.

The early Greek writers tell us of the famous city of Atlantis, westward of the Gates of Hercules (now known as the Strait of Gibraltar), and interesting articles have been composed by modern authors on the subject. But I will leave to others the tracing of any connection between that wonderful locality and Florida. It was in this direction, and its site remains undetermined.

In more modern times, the daring Scandinavian voyagers, from Norway and Iceland, visited Greenland and planted colonies there. They also sailed southward along the east coast of North America. This was about the close of the tenth century. There is, however, no evidence that they visted Florida.

Columbus, on his first voyage in search of a route to the Indies, discovered San Domingo, October 14th, 1492, and on a later voyage the coast of South America, but he had no idea of the existence of Florida. In 1497, Sebastian Cabot discovered North America, but did not proceed as far southward as Florida. In 1512 or 1513, Juan Ponce de Leon, having heard reports in the West Indies of a wonderful youth-restoring fountain, that would smooth the wrinkles of age and restore the whiteness, smoothness, vigor and agility of youth to the one who drank its waters, made diligent inquiry and learned that it lay to the westward. After an extended cruise he reached the unknown coast of Florida, March 27th, on Palm or Easter Sunday, landing near Fernandina, April 2d. May 1st, 1562, a colony of French Protestants, under command of Jean Ribaut, entered the St. Johns River. Another colony of Huguenots left France, April, 1564, landed at St. Augustine, but located at St. John's Bluff, on the St. Johns River. The next year the French colonists were massacred by the Spanish under Menendez. In 1567, Dominic de Gourges avenged this act by the destruction of the Catholics. May 30, 1539, Hernando De Soto landed at Tampa with a large force, sent his ships back to the West Indies, and marched through the interior of the country as far as Georgia, and thence to the westward in the search for gold and populous cities, which were not to be found.

Matters were then quiet in Florida for many years, the lithe, hardy and freedom-loving native Indians, who had occupied the country from time immemorial, being alienated by the cruelties of the Spaniards, successfully resisted them, and attempts to subdue and colonize the country were abandoned until it passed into the hands of the English, February 13, 1763. A few attempts at settlement were made, but abandoned in 1783, when it was re-ceded to Spain.

The United States made a treaty with Spain in 1819, making Florida a part of our domain, the transfer taking place July 10th, 1821. It then became a Territory, with General Andrew Jackson as Military Governor. May 3d, 1822, Congress established a civil government and William P. Duval was appointed Governor.

The Indians held the whole State, with the exception of a few military posts on the coast and along the northern border, and troubles with them continued frequent. While under Spanish rule it had become a resort for desperadoes, and those guilty of crime in the States to the North, which retarded development and tranquility. December 28, 1835, a fearful Indian war was commenced by the massacre of Colonel Dade and his force of 115 men, and a general uprising of the Indians to resist removal to the west of the Mississippi River, which had been determined upon by the General Government. It lasted until 1842, the Indians being subdued. This year the Armed Occupation Act was passed to hold the land by force.

March 3, 1845, the State was admitted to full membership in the Union. The State progressed slowly, Indian troubles from 1855 to 1859 keeping matters in an unsettled condition and retarding emigration and development. January 10, 1861, the State seceded and joined the Southern Confederacy. The war being over, the State returned to the Union, October 28, 1865. Since then the development has been very rapid, surely and steadily augmenting from year to year.

At first the immigrants were mostly from the Southern States, near the northern border, mostly from Georgia, though there have been increasing numbers from South Carolina, Alabama, Mississippi and other States.

The people of the North did not begin to take much interest in Florida until after 1875. The State had been a Spanish possession from the time of its discovery, in 1512 or 1513 (the exact year is unknown), until it passed into the possession of the United States, in 1821, with the exception of the English occupancy from 1763 to 1783; and where Spain rules, civilization languishes. Her power is the bane to progress. After this the Indian troubles kept things in a turmoil until all culminated in the civil war, which impoverished the whole South. Is it any wonder that Florida had made so little progress up to the opening of the present decade, although she has the most delicious climate in the world, a soil and productions that give the greatest and the most profitable returns to the acre of any State of the Union, a land where seed-time and harvest go hand in hand, where crops grow every month in the year, successive productions being taken from the same piece of ground; a land where the cultivator can work in his field every day in the year, if he chooses; a land where nearly every production and industry can thrive; where labor, energy and intelligence secure a sure reward; where one's needs, wants and necessities are less than in other portions of the country, and the necessaries of life more easily obtained; where provision does not have to be made for bitter and inclement winters; where the summers are

pleasant, comfortable and enjoyable; where the poor man can secure
a competence for himself and family, and the man of means increasing wealth and fortune; where bitter storms can never come; where
the general healthfulness is greater than that of any other State;
where acclimatizing fevers are unknown; where the most virulent
diseases of other sections, pneumonia, diphtheria, typhoid fever‘
phthisis, yellow fever, hydrophobia, sun-stroke and several other destructive ailments, are either unknown or very rare; where malaria is not
prevalent; where the occasional sickness is of a milder type than in
other States, and yields rapily to proper medical treatment; a land
where there is abundant and profitable employment; where beauteous
homes can be easily made, and where flowers bloom all the year?

The enumeration might be continued for pages. Such are a few
of the inducements offered by that choicest and most delectable portion of the State, Polk County. Great as may be the charms of other
portions of the State, and there are many, Polk County claims to surpass them all in desirability and attractiveness. It is the centre of
the most delicious and delightful zone the State affords. Her people
are not given to boasting and self-praise; they present the solid attractions, that those who see may understand. Polk County makes
no depreciative or invidious comparison with other sections. Their
attractions are great and genuine. Where there is a good there may
be a better, and Polk unhesitatingly claims the best. Come, see and
believe.

The northern part of the State has many attractions, and was the
first settled. The growth there has not been rapid. The climate, soil
and productions are essentially the same as those of the adjacent
Southern States. Being further south, the climate is somewhat milder,
but being a part of the continental portion of the United States, the
characteristics of the peninsula are absent. It has, however, a variety
of charms and attractions that will eventually ensure its development.

It was the fashion and the custom until quite recently to decry
South Florida. Parties interested in the northern, or continental part
of the State, represented it to be a vast swamp, interspersed with barren sand hills. They claimed that it was virtually uninhabitable by
man, being the home of alligators, poisonous serpents and other disagreeable vermin, and so malarious and insect-infested that civilized
man could not exist. Also, that the heat was too intense to be endured.
These phantoms of their selfish imaginations have been dispelled by
the light of knowledge derived from actual experience, and South
Florida is found to be the most delightful and the most healthful part
of the State. Its products are also found to be the most valuable and
the most reliable. As it becomes better and better known, immigration is pouring into this portion of the State much more rapidly than
into any other, and active towns and busy, bustling cities are springing up as if by magic, especially through the central portion from the
Atlantic to the Gulf. Railroads are also being built here very rapidly.
Everywhere energetic progress is manifest.

The Indian troubles that continued until 1859, not only prevented
any considerable settlement in this southern peninsula, but it also kept
the old-time planters from locating here with their slaves. Consequently this favored portion of Florida is not encumbered with the

flotsam and the *jetsam* left by the late war in the more northern portions of the State. South Florida is ruled by white men without any fear of negro supremacy. It does not need any peculiar provisions to be inserted in the State Constitution to prevent the intelligence and wealth of the community from being helplessly outnumbered and overridden by a large, ignorant, servile population with whom they have and can have no sympathy. This is the land of the free, not the slave.

It is only a decade since the tide of Northern emigration *began* to be attracted to Southern Florida. They reached it by the St. Johns River, landing at Sanford. Previous to this time the most of the little emigration to the State, and more especially to South Florida, was from the Southern States. About ten years ago there was a little spurt of Northern immigration. The majority spread out from Sanford, on Lake Monroe, the head of large steamboat navigation on the St. Johns River, throughout Orange County. A few stopped off in Volusia County, notably at DeLand. A few small and feeble colonies located on the Atlantic Coast, the colony from New Britain, Conn., locating at Ormond, on the Halifax River, being the most notable. Orange County made a beginning, but it was eighty-five miles through the primeval forest without roads or bridges to the capital of Polk County. Northern people did not wish to bury themselves in the woods away from all the privileges that civilization affords, and they did not come. Besides being unknown, this whole southern region was considered virtually uninhabitable.

The wheels of Time continued to revolve and 1880 appeared on its dials. The few newspapers, the occasional letters from the happy and prosperous settlers, and the testimony of the increasing number of prospectors, wielded a powerful influence in attracting the attention of the people of the whole North. The tide of emigration to the West had reached and overleaped the Rocky Mountains, and there was a rebound. They too began to turn their attention to Florida. What Western people do, they do on a grand scale, with vim and energetic push. They have less timid conservatism than the people of the East. They are broader and more liberal in their views. They are more ready to adapt themselves to changed conditions, to pull off their coats and go in to win.

Western as well as Eastern people are coming to South Florida. With their combined energy and means, they are making lovely homes, building thriving towns and busy, bustling cities in the tier of counties to the North of Polk; in Volusia, Orange, Sumter and Hillsboro' Counties. They have done well; their choice was a wise one, for there was neither railroad nor steamboat, nor other ready means of access to and from Polk County, and its more delicious climate and more fertile and productive lands. A few energetic pioneers from those sections pushed on to Polk and were not only surprised, but delighted with the intrinsic excellence and attractiveness of the country they saw.

The figures on Time's dial mark 1884, and a railroad from Lake Monroe to Tampa runs through the northern portion of Polk County, through a wonderful Lake Region. Thrifty towns sprang up along the line with a celerity only equalled by the wonders attendant on

Alladin's lamp. The South Florida Railroad was the lamp, and the Plant Investment Company the Alladin that caused this wonderful transformation in the primeval wilderness.

Now 1885 appears and the railroad is opened into the heart of the county, to Bartow, with probabilities of speedy extension. The Florida Southern, running through the interior of the State, from north to south, forms a junction with the South Florida at Lakeland, and is soon to be extended to Bartow and beyond to the south. Rumors of contemplated railroads fill the air. People from all sections are coming to see the country and are delighted. Their friends, neighbors and acquaintances follow them as soon as they can arrange their affairs at their present homes. Everywhere is surprising activity, land cleared and buildings erected. The climate is the most delightful throughout the year yet known, the lands are fertile and productive, the views are charming and general good health prevails.

As regards general beauty and attractiveness, Polk County is virtually a continuation of Orange County into a more southern latitude, where the winters are milder and the summers cooler. Like Orange it has immense numbers of beautiful lakes, and they are well stocked with fish. There are, however, many more running streams and beautiful valleys ; the ground has a firmer and more substantial tread, the soil is more fertile, and oak growths are more abundant and vigorous, which, with its varied surface, gives pleasing additional attractions to the landscape. I would draw no invidious comparisons with Orange, the banner county of the State and my adopted home, for Orange County deserves every line of praise that has been given her. I know no county in the State that is, at present, the peer of Orange, and believe there will be none excepting Polk, which I judge to have the most varied, desirable and magnificent natural charms and attractions of any section of fair Florida, the future home and resort of millions of the world's energetic and enthusiastic, as well as tired, weary workers. To these Polk County will be a veritable HOMELAND. Polk County, like Orange, is an interior county, a watershed, and the balmy and refreshing breezes from Ocean and Gulf, changing semidaily, are filtrated through miles of balsamic pine on either hand before sweeping over her healthful and inviting territory. Polk has all the latter's natural excellence as well as a number of additional charms and advantages.

Polk County has had to await the apparently slow, yet really rapid, processes of settlement and development, without either natural or artificial highways of ingress or egress. The counties to the North, from Polk to the city of Jacksonville, in Duval, have had the magnificent St. Johns River and fleets of steamers to aid in their development, while, until the present year, Polk County has not only had no steamers, though there is a river running through it from North to South, but Congress would appropriate nothing for its improvement because it was called Peace Creek. The Legislature has changed its name to Peace River, and an appropriation will doubtless be forthcoming, that Polk may have a navigable outlet to the Gulf. Kissimmee River forms its Eastern boundary and the Drainage Company has recently rendered that navigable. Hence, as Polk County is becoming known, her water ways are being made ready for use. Canals could also be easily cut, and thus her magnificent lakes could be utilized for transportation.

I take great pride in the glorious development of Orange County during the past five years, since the commencement of her railroads in 1880, feeling that I have contributed my mite thereto, and would add to rather than diminish one word I have said in her praise. Her population and wealth have quadrupled, her citizens are the *creme de la creme* of the whole country ; the number of buildings erected and the business industries developing there are astonishing and her wild lands have increased in value from $1.25 to $25, to $50, to $100 per acre, while near the numerous thrifty, fast growing towns $200 to $500, and even $1,000 per acre, is no uncommon price. Even this progress is surpassed in the most wonderful city of Orlando, the county seat, where (I quote an editorial in the *Reporter* of that city) business lots sell at the rate of thirty-six thousand dollars ($36,000) per acre. Yet three to four years ago the pine woods were growing over nearly all the present site of the city. I know this to be a fact for I was there.

There seems to be no possible reason why the development of Polk County should not be equally great and even more rapid. I have referred on previous pages to her advantages and attractions. Polk County was much more developed when the first railroad reached her county seat, last January, than was Orange when a like event occurred, and five years will make even greater changes here than the same period has there, wonderful and surprising as the developments have been.

Bartow has abundant room for expansion in all directions, and especially to the South, along one of the very finest and most desirable plateaus to be found in any State. Already dwellings and beautiful groves are quite numerous and rapidly increasing. Soon the whole twelve miles from Bartow to Fort Meade will be one continuous village. The prices of lands are very moderate yet, but the rush of emigration is assuming such immense proportions that prices are sure to advance very rapidly. There is no mistaking the determination of the people of the other States throughout the Union to immigrate from the regions of terrific storms and destructively cold and severe winters. This is evidenced not only by the numbers already settling and prospecting, but also by the interest manifested everywhere, even across the Atlantic, in Florida, and more especially in South Florida.

The increase of values has been, and is, so great that many who had purchased larger tracts than they really needed, so much less land being required here than in other sections, were able in a short time to pay for the whole by selling a part, so greatly had the value advanced. The same will be the experience of thousands of others who invest here before prices get beyond their means.

POPULATION.

In 1850, the whole county was virtually untrodden by the foot of the white man, save at intervals by an adventurous hunter, though there were a few, and but a very few, log cabins to be found among the head waters of the Alafia and Hillsboro' Rivers. Even the few

adventurous pioneers who located there sought safer quarters upon
the outbreak of the Indian troubles in 1854, consequently 1860 found
a very limited number of settlers in the county, and they had not
even a court-house, nor was the site determined until 1866, when the
county building was erected at Bartow. Nor was there a church
building or a store, and none but the most primitive of school houses.
By the census of 1870 the population is given as 3,113, but the number
is believed to be largely exaggerated, as the reliable census of 1880
gave but 3,156 population. These figures are largely increased by
immigration, during the past five years, from all parts of the Union,
the estimated present population being about 7,000, having doubled
in five years. A census will be taken during the summer, which will
give the exact figures, and be published in the Trade Edition of the
Times-Union in October. From this time the population is sure to
double very rapidly.

REAL AND PERSONAL PROPERTY.

By reference to the tax books of 1866, I find 132 white males
over 21 and under 55 years old, and 80 colored. The 13,752 acres of
and owned by individuals was valued at $50,104; the capital in trade,
$4,000. There were 352 horses and mules, valued at $33,975. The
cattle, hogs, etc., numbered 37,696, with an assessed value of $178,174.
There were 272 sheep and goats, valued at $716. The carriages,
carts and wagons numbered 135, and were valued at $4,270.
Household furniture was valued at $8,505; musical instruments,
jewelry, etc., at $485. Total valuation, $325,604. Tax assessed,
$2,024.02; collected, $1,920.67. The poll-tax was $3 each.
In 1870, the number of polls has increased to 259, and the tax is
reduced from $3 each poll to 50 cents. The land assessed is 15,117
acres, with a valuation of $19,133, Of this land 1,943 acres are im-
proved, or cultivated, the value of the improvements being $23,051.
There are 423 horses, 1 ass, 47 mules, 28,401 neat and stock cattle,
189 sheep and goats, and 3,684 swine. The valuation of household
furniture, etc., is $105,627, and the full cash value of personal property
is $303, 489. The State tax was $1,745.14; county tax, $872.57, and
the school tax, $698.05.
In 1880, the number of acres assessed have increased to 29,022;
acres improved, 1,535; valuation, $70,090; valuation of town lots and
improvements, $13,590; horses and mules, 655; cattle, 22,030; sheep
and goats, 554; swine, 6,330; full cash value of animals, $156,234;
value of personal property, except animals, $69,343; aggregate value,
$225,577; while of real estate and personal property it is $309,257.
Total state tax, $2,375.30; total county tax, 2,529.93, of which the
general school tax was $309.26; county schools, $773.14. Males over
21 and under 55 are 421.
In 1881, the assessed valuation was $368,870; in 1882, $684,702;
in 1883, $1.036,223; in 1884 it had increased to $1,689,053. During
the past year the increase has been very much greater than in any
previous year, as is shown by the assessment now in preparation. A
number of new and substantially built towns, with stores, post offices,
churches, schools, telegraph and other offices, railroad depots, and the

varied appliances of energetic civilization having taken the place of the primeval forest.

In 1884, we find the number over 21 and under 55 to have increased from the 212 of 1866 to 670; the 13,752 acres of assessed land in 1866 have increased to 223,196 acres in 1884; the improved to 7,288 acres. The value of the land is $1,122,758; of town lots, $59,025. The number of horses, asses and mules is 950; cattle, 24,-108; sheep and goats, 1,651; hogs, 7,134, with $225,421 as the cash value, and $279,149 as the value of other personalty, making a total of $504,570 for personal and $1,686,373 of both real and personal property. The general revenue tax of three mills yields $5,405.91; the school tax of one mill gives $1,420.73; making the total state tax $7,092.94. The county tax proper is $4,560.98; school, $5,059.91; special, $1,688.73; for court-house, $6,748.94; making a total for county proper of $18,056.86; it being two and a half mills; while the total tax of the county for 1884 was $25,149.80. The present year shows a very large increase of improvements and of resources of all kinds.

SCHOOLS.

Very greatly increased interest is being shown in educational matters. The number of pupils of lawful age is about 2,000. There are fifty-three schools organized, of which two are colored. Bartow, Fort Meade and Lakeland employ assistant teachers. The last two have just erected fine new school buildings and Bartow will soon invest about $20,000 in supplying her needs in the matter of educational facilities.

ORANGE GROVES.

The number and acreage of orange groves has increased immensely with the increase of transportation facilities, having probably doubled within the past two years. At Bartow, the most reliable estimates give not less than 12,000 trees in grove form, with some millions, from one to five years old, in nursery. The most of the trees in the groves are set thirty feet apart, though many are set at a less distance, thus increasing the number, as my estimate is for fifty trees per acre. At Fort Meade there are about four hundred acres in grove, or some 20,000 trees, while there are millions in nursery. Here, too, the amount of land being set to orange trees is increasing with wonderful rapidity. A trip over the twelve miles between Bartow and Fort Meade shows a very large acreage of groves that are either in bearing or are making astonishing progress. They are of all ages. There are also numerous groves, not only near other prominent centres, like Lakeland, Acton, Auburndale, Winter Haven, etc., but also throughout the whole extent of this climatically-favored county, where much of the soil is so fertile that trees reach maturity without fertilization and also yield good crops, year after year, to the cultivator. Many groves are in bearing and give immense returns of delicious golden fruit to their happy and prosperous owners.

A WORD OF ADVICE.

To the thousands whose thoughts are turned towards Florida, I would say, be very careful to consider South Florida by itself. Do not fail to make the distinction necessary between South Florida, and especially between Polk County, and the State at large. Every portion has its own especial attractions, advantages, comforts and discomforts. Each section should be judged by itself and stand or fall on its own merits or demerits. Do not forget that Florida is a *very* large State. Do not fail to remember that though the whole four hundred miles from north to south, bears the name of Florida, the climate and productions are very much diverse from those of any equal extent of territory in the North, the great West, or the South. Do not fail to give due weight to the fact that SOUTH FLORIDA is a peninsula; that the Gulf Stream flows along its western, southern and eastern coasts, modifying its climate, its temperature and its productions. Do not fail to remember that the delicious and invigorating benefits derived from the trade winds are only to be secured in South Florida. Remember that the life-giving, heat-dispelling breezes of the day and the soothing, sleep-inducing, strength-resting coolness of the nights are only to be secured in their full perfection in South Florida, of which Polk County is the most desirable and advantageous centre.

Do you fear troublesome insects or reptiles, remember that Polk County is centrally situated, and that its elevation above the sea and the absence of salt or other marshes gives it an exemption unkown on the coast. In many of the higher and more prominent localities mosquito bars are unknown. In fact, it is more exempt from insects and other pests than the average localities of the Union.

Especially do not group all the eight distinct belts or zones of Florida as one homogenous whole, as their characteristics are very decidedly varied, and their productions essentially different. Polk County is decidedly semi-tropical, and its productions very, very different from other parts of the country. Those who come here have to unlearn the lessons they have learned in other parts of the Union and begin anew if they would succeed.

If you are so situated that you are perfectly and undeniably contented where you are, do not come to Polk County.

If you cannot tear yourself away from old associations and form new, do not come to Polk or any other county in Florida. This land has been reserved by Providence for those who desire to renew shattered health or fortunes, as thousands are doing, and for a happy and delightful abode.

The people here are as good, as kind, as sociable, and as generous to their acquaintances or to strangers as those of any country in the world. There is no difficulty in securing pleasant homes here. In fact, there is less rancor in politics or in religion than in any other part of the world. The people desire increased immigration, especially of educated, refined and well-to-do people, as that insures more and better schools, churches, stores, roads, cultivated fields and improvements generally that appertain to the highest civilization. Everything is progressive here. The capacities and capabilities of this section are wonderful, but their development has hardly commenced.

If you desire to practice farming, as at the North or West, stay there. Though corn can be raised here and sold at a price that makes it more profitable than at the far West, this is not an agricultural, but a fruit and vegetable country. The farms must here be made gardens or orchards, and the intensive system of cultivation must prevail to secure the best results. Larger returns are secured from five acres here than from fifty at the North. Consequently, though large holdings will give a great profit from the rapidly-increasing values of land, a few acres give all that is needed for the purposes of cultivation. It is simply a question of large returns from a few acres here or small income from each of many acres in other sections.

You can buy a through ticket to Bartow, Polk County, Florida, at any of the leading railroad offices of the country. This will take you all the way by rail, or you can go by steamer from the principal Eastern cities to Fernandina, and thence by rail to Bartow, or you can make the trip from Jacksonville to Sanford by steamer up the St. Johns River, and the remainder of the trip by the South Florida Railroad.

Men have come without moneyed resources and have prospered, securing wealth by their own labor and enterprise, but the majority of people will find it more pleasant to be provided with a reasonable . amount of cash ready for use, in case they are pleased with the opportunities for investment.

Do not expect to find things here altogether the same as in the place from which you came. If they were, you might as well stay at home. Climate, soil and vegetation, all, are different from the Northern or Western country. Difference in lattitude produces a difference in the habits of the people.

Here you can comfortably pass the most of your time in the open air, hence, if the hugging of a base-burner coal stove and the breathing of a close and poisonous atmosphere be your chief delight, remain where they abound, for such have no place in South Florida. Here the sun gives the requisite amount of heat, and the delicious breezes supply an abundance of pure and wholesome air to breathe.

If you should be so fortunate as to come to Polk County, do not be so egotistical as to think you know more about the country and the way things should be done than those who have been here for years and obtained their knowledge by experience. However smart and wise you may esteem yourself, it will be well for you to remember that the people here are your peers at least. They not only have a practical knowledge, but many of them are quite as highly educated and cultured as yourself.

In the villages you will find as good and as intelligent society as in any part of the Union, and in the country, as kindly neighbors as are to be found anywhere; nor will you be farther from them than in other parts of the country at large.

Do not be in too much of a hurry at first, but settle down quietly in some pleasant locality, rest from your journey and make yourself acquainted with the characteristics of things about you.

Do not locate on land that is flooded at times of high water. It may be cheaper at first, but it will prove more expensive, as well as disagreeable, after. There is plenty of good land, but there is much more that is undesirable.

All who propose to make a stay of a few, or many years, in South Florida, should purchase land and improve at least a portion of it.

Land here is rapidly increasing in value, and will continue so to do
until the minimum price of the best lands will be not less than $1,000
per acre. Those with groves will be worth much more, while in towns
and cities the value will be rated by tens of thousands of dollars.

The rapid advance in the prices of land in South Florida is not due
to a speculative boom, but to its substantial development. The advance
has been steady from year to year with the increase of population,
and of the facilities for railroad transportation.

The new comer requires more cash capital now than a few
years since for several reasons; the lands are higher-priced, and consequently more money is required to secure a place; people also build
much better houses, as a rule, and also live and dress more expensively;
there are more of the appliances of what is called modern civilization.
Those who choose can rough it, but it is difficult to get far away
from neighbors.

If one comes here with the design to work for others—and there
is much work being done—he should also secure at least a few acres
of land and make it his savings bank and donate his spare time, his
dimes and dollars to its improvement. He will thus, in a few years,
become independent.

Women feel the change from an old country to a new more than
men, as it is more difficult for them to form new associations, but as they
are credited with the possession of more tact, patience, self-denial and
self-sacrifice, they should be able to endure the breaking of old ties,
that the fortunes of the family may be permanently bettered.

Ladies who delight in flowers can have their yards filled with the
most beautiful throughout the year, there being no destructive freezes
here, and but few, if any, frosts. They can enjoy a wealth of bloom
throughout the year.

Here one can live in cottage or mansion, as means or taste may
dictate. Social distinction is not based upon wealth, but upon energy,
intelligence and true and desirable qualities of heart and mind.

THE POLITICAL SITUATION.

You need not fear being ostracised should you be a conscientious
Republican. There does not seem to be half the bitterness of party
feeling here that there is in other parts of the country. There are and
have been, but very few negroes here, consequently the race bitterness
of some sections of the South is unknown. This is a "white man's
country," and it is also becoming quite cosmopolitan, hence, the
question is, "What kind of a man are you?" not "What is your politics?" Speech is as free here as at the North or West, but people
come here to make orange groves, to retrieve wasted health, to make
delightful homes, and to secure either competence or fortunes, not to
dabble in political cess-pools. The older residents are mostly Demo-
crats, the later arrivals represent all parties, while the tendency is for
the better elements to work together to secure the greatest possible
good for this section of the country. There is no surveilance of or
disturbance at the polls. Election days pass off quietly, and as the
ballot is cast so is it counted.

THE SOUTH FLORIDA RAILROAD.

Leaving the main line of the South Florida Railroad at Bartow Junction, the tourist passes over the Bartow Branch, 17 miles, to Bartow, the county site of Polk County, one of the most beautiful towns in South Florida.

The situation of Bartow and the surrounding country is such as to give promise of a large increase in population, in business and in importance in the near future. The connection made by the railroad line with Tampa on the West coast and Sanford on the St. Johns River, has added very materially to the prosperity of the town, and where only a few years ago wild land was found are now to be seen charming dwellings, beautifully located and surrounded by flourishing orange groves.

The South Florida Railroad, in every way consistent with business principles, develops the country through which it passes and gives to the new settler all advantages to add to his income by prompt delivery of produce at the Northern markets, and keeps producers well informed as to market values by constant telegraphic reports from New York, which are posted three times weekly at prominent places in towns along its line.

The Bartow Branch passes through some of the most beautiful and fertile land in South Florida—through the centre of the noted Lake Region of Polk County. The road passes along on the summit of a ridge, on both sides of which bright lakes are seen, on whose shores are found a succession of orange groves and gardens.

The soil around Bartow is particularly adapted to vegetable culture, and the departure of trains from that point is such as to enable the producers to put their vegetables into a Northern market in the shortest possible time. Rail communication is soon to be opened over the main line of the South Florida Railroad by a branch to the North, connecting with the Florida Southern and then on to the point of connection of that road with the Savannah, Florida and Western. This will materially reduce the time from Bartow and make the place as easily accessible to the tourist, business man or invalid as has been Jacksonville in the past years.

The lines of the South Florida Telegraph Company extend from Bartow to all points on the line of the South Florida Railroad and its branches, and at several of these points connects with the wires of other companies, making direct telegraphic communication between Bartow and all points in the North, South and West, as well as to points in Cuba.

Winter Haven, one of the most beautiful spots in Polk County, is situated 5 miles from Bartow Junction, and 12 miles northeast of Bartow, surrounded by lakes. One can there feel almost safe from frosts, as the cold winds are tempered by the warmer water over which they pass. This town, while started only a few months since, is already a prosperous and growing community and gives promise of increased prosperity in the future.

Trains leave Tampa and Sanford both morning and evening for Bartow and points on the Bartow Branch. The tourist and hunter will find no more delightful country than that through which the Bartow Branch of the South Florida Railroad passes, being unequalled in scenery, and affording to the sportsman all kinds of game and in the lakes a variety of fish unknown outside of Florida waters. First-class passenger accommodations are furnished by the railroad. parlor cars over the main line, air brakes, steel rails, smooth roadbed and all the comforts that can be found on the large Northern roads.

Tickets for Bartow can be procured at all the principal points in the North, East or West, at the Savannah, Florida and Western ticket office in Jacksonville, or on boats of Peoples' Line of steamers.

For further information, apply to Frederic H. Rand, General Freight and Passenger Agent.

OFFICIAL DIRECTORY

OF

POLK COUNTY, FLORIDA.

CIRCUIT COURT—Sixth Judicial Circuit—H. L. Mitchell, Judge ; S. M. Sparkman, State Attorney.

COUNTY COMMISSIONERS—Col. J. N. Hooker, Chairman ; J. F. Kelley, J. H. Kirkland, N. B. Norton, B. F. Holland.

COUNTY JUDGE—James A. Fortner, Bartow.

COUNTY CLERK—William H. Johnson, Bartow.

BOARD OF PUBLIC INSTRUCTION—M. D. L. Mayo, Chairman ; J. T. Wilson, J. W. Brandon.

SUPERINTENDENT OF SCHOOLS—John Snoddy, Bartow.

COUNTY SURVEYOR—J. W. Boyd, Bartow.

COUNTY ASSESSOR—U. A. Lightsey, Fort Meade.

COUNTY COLLECTOR—J. B. Tillis, Fort Meade.

COUNTY TREASURER—F. F. Beville, Bartow.

SHERIFF—R. T. Kilpatrick, Bartow.

NEWSPAPERS.

Bartow Informant, Bartow; G. A. Hanson, Editor ; D. W. D. Boully, Publisher.

Lakeland News, Lakeland ; L. M. Ballard, Editor and Publisher.

Fort Meade Pioneer, Fort Meade; F. Q. Crawford, Editor and Publisher.

JUSTICES OF THE PEACE.

Precinct No. 1—D. C. Lancaster, H. E. Padgette, Chicora P. O.

Precinct No. 2—V. L. Tillis, Fort Meade.

Precinct No. 3—George S. Durrance, Bartow.

Precinct No. 4—R. E. Windham, Medulla.

Precinct No. 5—J. W. Tucker, Lakeland.

Precinct No. 6—William L. Patterson, Sanitaria.

Precinct No. 9—J. A. Fortner, Bartow.

Precinct No. 10—Eppes Tucker, Lakeland.

DESCRIPTIONS OF LOCALITIES.

BARTOW.

Prominent among the energetic, pushing, beautiful and rapidly growing towns of South Florida is Bartow, the central and capital town of far-famed Polk County. It is situated near the centre of the elevated and fertile ridge of land, the backbone of the peninsula, that extends in a northerly and southerly direction until lost in the flat lands of the counties in the extreme southern portion of the State: Here, however, the road-bed of the South Florida Railroad has an elevation of 114 feet.

At and surrounding Bartow, on every side, are large quantities of the most fertile and the most desirable lands that are to be found anywhere in the State of Florida. The vigorous growth of the beautiful pine, intermingled with an abundance of oak of a number of varieties, is a surprise to the visitor who had become tired and wearied of the ceaseless, unbroken pine presented by so many sections. There is a charm in variety, and a fertile soil has an undeniable attractiveness.

Another surprise that arrests the attention of the visitor, is the firm and solid tread that greets the impress of the foot of man and beast. No wading through deep and difficult sands, but hard, smooth and enduring pathways. There is also a variety of surface, which is gently rolling, precluding the monotony caused by broad stretches of flat lands. Here are hill and dale, with gentle swells, furnishing delightful building sites and ample drainage—the central portion of the town being higher than the surrounding and contiguous country.

The broad streets and avenues cross each other at right-angles, the blocks being of one acre each. Along the streets and in the yards of the residents are numerous vigorous oaks and other trees that give a most delightful and congenial shade, the like of which is to be found in but few places outside of the fertile ridge of Polk County. Here, too, are vigorous orange trees, laden with an immense quantity of the "apples of the Hesperides—fit food for the gods."

The centre of attraction, and of business, is the capacious, elegantly and substantially furnished court-house, the finest in South Florida, which occupies not only a central, but also the highest, part of the town. The fence around it encloses a square of one acre of land—one block—which is surrounded by a line of posts connected by chain cables. About a rod within is a substantial picket fence. On the four streets surrounding, and their extensions, are grouped many of the stores, offices, hotels and business houses.

The depot of the South Florida Railroad is situated about half a mile to the southeast of the court-house which is thus secure from the annoyance caused by the noise of trains, the roar, jostle and push of active business. The court-house fixes the centre, and new buildings, mostly of a very neat, elegant or substantial character, are being erected in every direction around it. Evidently, judging from the present rate of progress, it will be but very few years before the whole of the four miles comprising the area of the corporation of Bartow will be completely covered with buildings and groves, from sixty to seventy new buildings having been erected and many acres set in orange groves since the advent of the South Florida Railroad, last January, while work on both is progressing very rapidly.

A careful computation shows that about one-sixth of the area of the corporation, some four hundred acres or more, are already occupied by buildings, groves, nurseries, etc. Lemons, limes, guavas, bananas, grape fruit, Japan plums and persimmons, strawberries and a long list of other fruits are cultivated, as well as oranges, besides the beautiful shrubs, flowers, running vines, etc., etc., that adorn the yards of so many of the residents. This is the true HOMELAND, where meet the productions of both the temperate and the semi-tropic zones, where harvests are continuous and where flowers bloom all the year, destructive freezes being almost, or virtually, unknown, and frosts rare and mild; its comparatively low latitude giving it much greater exemption from these destructive influences than localities at any distance to the north.

The survey of the Florida Southern Railroad runs within less than a quarter of a mile to the west of the court-house, and work is progressing rapidly; a line already building from Tavares to Charlotte Harbor is to run through Bartow; a road is chartered from Bartow to Tampa, and other roads are expected to link Bartow with other parts of the State in all desirable directions, making it a lively railroad and general business centre. In fact, its future seems to be indisputably assured as one of the most important and desirable business centres of South Florida.

Bartow is the natural business centre of a large extent of the most fertile country, as well as the most healthful and salubrious and desirable, that any part of Florida affords. The water, too, is excellent, and is readily secured by boring or digging through alternate strata of sand and clay to a depth of 25 to 30 feet, at which depth the supply is constant and unlimited. There is also a surprising and unexpected exemption from insect pests, mosquitoes being few in number and so rarely seen that bars are unused and unnecessary. Flies, gnats, etc., are also limited in quantity and fleas are disappearing as the laws banish the hogs.

The prices of lands are very reasonable, considering Bartow's importance as a trade centre; the rapidity with which the town is building in every direction; the prospects as regards railroad facilities; the favorable location for a variety of manufactories; the fertility of the soil; the delicious healthfulness of the climate; the large quantity of choice outlying lands; the great variety of fruits and vegetables that can be successfully and profitably raised and marketed; the many beautiful and desirable locations for homes and for business

places; the social and enterprising character of the people and many
other reasons that will suggest themselves to the visitor.

The first settlement in the corporation's limits was made in 1851.
In 1852 several families settled in the near vicinity. Being far distant from transportation and without good roads—Tampa, on the Gulf
of Mexico, forty-five miles distant, being the nearest trading post and
post-office—the population increased very slowly, notwithstanding the
remarkable fertility of the soil and delicious salubrity of the climate.
In 1866, Bartow was made the county seat and the International
Ocean Telegraph line was built, and opened an office here. The first
store was built and opened the same year. The court-house, a hotel, a
school-house and Masonic lodge, and several other buildings, were also
erected in 1866. Then things resumed their usual quiet course, the
lack of transportation being an insurmountable obstacle.

The population increased very slowly and no attempt was made
to build a town. The chief industry of the people was the raising of
cattle and agricultural products for home use. The people were self-
supporting from the fertile soil. In 1868, Capt. David Hughes located
here, built a store-house, and went into the cattle business on a large
scale. W. T. Carpenter had the first and only store for the sale of
goods, from early in 1865 to 1870, when Capt. Hughes opened his
store to the public and has since done an immense trade.

Thus matters continued until 1881, in a quiet humdrum way, the
people being virtually isolated from the outside world. They had
plenty on which to live, but little else, on account of the lack of markets and the difficulties and expense of transportation. They necessarily became self-reliant ; they were happy and contented ; crime was
very rare. Railroads were chartered occasionally, but until 1880 none
were built in South Florida. That year the South Florida Railroad
was built from Sanford to Orlando, and extended in 1881 to Kissimmee.
It was chartered to run through Bartow to Tampa. Then railroad
talk became rife ; a few enterprising prospectors scoured Polk County
searching for desirable lands, and brought back glowing reports of the
beauty, fertility and delicious salubrity of the country. They dwelt
with enthusiasm upon the rich lands, the vigorous growths of oak and
pine and other woods, the running streams, the rolling country, the
pleasant vales, the lovely building sites, the inexpressible deliciousness
of the climate, and the wonderful opportunities to secure fortunes.

G. W. Smith, one of Bartow's most enterprising citizens and
prominent merchants, came in the spring of 1881 ; the trip from Orlando, with his family and household effects, being made in ox-carts
over rough trails and swollen unbridged streams. They were eight days
on the way, camping at night. Is it any wonder that Bartow, or Polk
County, fertile and delicious section as it is, is not more thickly
peopled, or that now the South Florida Railroad has its present terminus at Bartow, people should be rapidly pushing into the country
to secure homes ? Mr. Smith was pleased with the country, its advantages and its opportunities, and having had extensive experience in
other sections, he knew a good thing when he saw it. He therefore
purchased about one-sixth of the then surveyed town, bought a saw-
mill, and later opened a store, to which he has made successive addi-
tions to accommodate his steadily increasing business. During the

year he proposes to build a still larger and elegant store, though his present place holds an immense stock of goods.

But I have neither the time nor the space to trace the individual history of Bartow. The impetus it received in 1881 has been earnestly progressive. July 1, 1882, the corporate government was organized, J. H. Humphries, Esq., the present Polk County delegate to the Constitutional Convention, being elected Mayor. Only twenty-eight legal voters were found in the corporation limits, and of these twenty-two were present. They knew the time had come for Bartow to move. The population is now about 700.

Numbers of new buildings were erected during the season of 1882 and 1883. Prominent among them were a Baptist and Methodist church, as well as stores and dwellings. Since that time building has been steady, progressive and continuous. In 1884, the old court-house was removed and the present fine and attractive structure erected. Also a fine hotel, an opera-house, and numbers of stores, residences, etc. In January of the present year, the South Florida Railroad reached Bartow, infusing new life into town and country. Its projection, its survey, and building to Tampa, had given a great impetus to the purchase of land and the setting out of groves, but an actual railroad here not only gave people the opportunity to come and see for themselves, but it also greatly facilitated and cheapened the transportation of supplies of all kinds. The manner in which its approach, even, gave an impetus to business and enterprise is evidenced by the fact that of the 12,000 orange trees set in grove, in the corporation limits, about half has been the work of the past two years. This fact alone indicates whether the many new comers have been pleased with the advantages offered by this section to clear-headed and enterprising men. Here many have builded their fortunes anew, many have made for themselves lovely homes.

The Bartow of to-day comprises a beautiful tract of rolling country, wide streets crossing each other at right angles, beautiful oaks and other delightful shade trees scattered throughout the corporation, substantial plank sidewalks and crossings in the chief business portion of the town, though the ground is so firm, and yet absorbs the falling rain so quickly, that they are very much less needed than in other sections of the country. The untraveled parts of the streets, the uncultivated parts of the yards and the fields are covered with a vigorous growth of grass. Orange groves abound, and outside of the business part of the rapidly-growing town the lessening tracts of pine, and of pine intermingled with oak, are patiently awaiting their destiny, for they will soon be removed to make place for buildings and groves, and gardens of fruits, vegetables and flowers.

To note the town itself, the large two-story court-house, with its tower and four gables, is an appropriate starting point. On the corner of Main Street and Broadway Avenue, to the south, is the general merchandise store of CAPT. DAVID HUGHES, well stocked with every variety of general merchandise. which is sold by his gentlemanly and attentive clerks at very satisfactory prices. The stock is not only large, but the yearly sales are immense, some years reaching as high figures as $60,000. Two additions have been made to the store since it was first built, to make room for the constantly increasing stock, attractively displayed, and last spring the active and wide-awake Captain, a Colonel and Commandant of the Militia of Polk County by commission, purchased the handsome and commodious opera-house and removed the clothing and gentlemen's furnishing goods to the ground floor, whereon is one of the most elegant and commodious stores in town. Yet his old store is full to overflowing. In front of Captain Hughes' store is a row of vigorous sour orange trees, protected by heaps of stone at their base.

Westerly from the court-house, on Broadway, is the general merchandise store of GEORGE W. SMITH, the pioneer merchant of this decade. His stock of goods is varied and extensive, but though his building is large and commodious he is unable to give them anything like an appropriate display. He will soon remedy this, however, by the erection of a new and creditable store, where he will continue his present immense business.

The LANG BROTHERS, on Broadway, directly west of the court-house, have a very fine and attractive stock of dry goods, clothing, gentlemen's furnishing goods, and boots and shoes, as well as a choice selected stock of staple and fancy groceries, canned goods, tobaccos, etc. The store is kept as neat as a parlor, and the goods are displayed in excellent style, while their prices are very moderate and encouragingly satisfactory. Their customers receive the most polite and gentlemanly attention, and those who once patronize them are sure to go again.

COLONEL J. N. HOOKER & Co's fine and well-stocked general merchandise store, on Main Street, to the southeast of the court-house, deserves a more than passing notice. It is not only an extensive, neat and well-lighted establishment, but contains a very heavy stock of general merchandise, embracing every variety, which are sold at prices to suit the times, by his attentive and gentlemanly clerks. The Colonel is Chairman of the Board of County Commissioners. He also has a large general merchandise store at Fort Meade.

Northeast of the court-house is the extensive hardware store of the REED BROTHERS, solidly packed with the great variety of articles in general demand, in the way of stoves, plows, pumps, piping, etc., etc. In fact, a variety of general hardware goods and general field and household articles too numerous to mention. They also do a general tinning business, drive and bore wells, etc., etc.

North of the court-house, we observe a floor laid beneath the shade of some handsome water-oaks, and supplied with seats. This is a first introduction to the BARTOW FURNITURE STORE, which, located just across the sidewalk, has an immense stock of the varied kinds of furniture most in demand. The goods are so numerous and so closely packed and piled that you can hardly move around, but you can, no doubt, secure the articles you desire.

J. P. STATHAM & Co. are enterprising druggists and physicians, located on Broadway. They have a varied assortment of druggist's goods, and are doing a popular and very lively business, when the exceedingly healthful state of the country is taken into consideration.

BAEUMEL & OPPENHEIMER, on Main Street, south of the court-house, are the new druggists from the West, who have built and opened a nice drug-store the present season. Everything is new and very neat and attractive. They also have a handsome soda-water fountain, and dispose of immense quantities of the cooling fluid. They set the example of self-protection from fire by means of a bored well in a rear corner of their store, to which a force-pump is attached. Water is forced to a tank in the attic, from whence, by pipes and hose, it is available in all parts of the building. They also have a Babcock fire extinguisher.

L. LYTLE has extensive livery, feed and sale stables, just to the east, on Main Street. He also deals extensively in carriages, hay and grain. He keeps a good supply of fine animals and carriages, and can insure any one a pleasant drive. Mr. L. is the pioneer livery man of Bartow, and does an immense business.

H. T. DIAL has a very extensive steam planing mill near the north edge of the town plat, and a saw-mill at Peace River, thus insuring a constant supply of lumber at satisfactory prices. He also has wood-working attachments, whereby he fills orders for orange boxes, vegetable crates, brackets, mouldings, etc. He also has a grist mill, and is contemplating starting a furniture manufactory.

J. M. DILL is the active and energetic contractor and builder, who is making his progress along the pathway of time, by the erection of substantial and creditable buildings. The work that he has done is his best and most convincing advertisement, and a bright future spurs him to earnest endeavor.

MRS. SNODDY's millinery store speaks for itself, and shows that there are some advantages in this direction, but a lady at one short visit would have more actual knowledge of the facts than a man could evolve in a week.

The three leading hotels of Bartow, taken in the order of the age of the buildings, are the WEBSTER HOUSE on Main Street, west of the court-house, E. Webster, proprietor; the CENTRAL HOUSE to the east, kept by J. F. Kelly, and the BARTOW HOUSE on Davidson Street, northwest of the court-house, by Dr. R. H. Huddleston. The rates at each are two dollars ($2) a day, the houses present a creditable appearance, and the proprietors apparently use their best endeavor to promote the satisfaction of their guests. The Webster House is being enlarged by a handsome two-story front.

TIGNER & TATUM, real estate agents, have their office on Main Street, directly south of the court-house. They are wide-awake and reliable men, thoroughly informed by years of personal experience, with the varied qualities and values of lands—past, present and prospective. They have large quantities of lands on their books, both improved and unimproved, and can suit their customers with town lots, bearing groves, pleasant and desirable residence lots, or wild lands in quantity, as the taste of the purchaser, or the condition of his pocket-book may dictate. They have been residents of Bartow and in active business here for several years, but had previously become well acquainted with other parts of the country, consequently they know land when they see it. As they have every kind of land for sale, they have no occasion to misrepresent the desirability of any particular tract, and their honorable reputation is good evidence that they have no such disposition. They have full faith in the future of South Florida, and especially of Polk County, and they have good and sub-

stantial reasons for the faith that is in them. They reply promptly
to all inquiries with regard to lands and opportunities for investment.
JOHN C. WRIGHT has an extensive general merchandise store on
the corner of Main Street and Broadway, to the southwest of the court-
house. He also deals largely in paints, oils, etc., doing an extensive
business in all lines. F. D. Beville, one of the early merchants of the
town, as it was commencing its later growth, is his chief clerk.
J. J. McKINNEY has a pleasant and well-stocked livery and sale
stable, a block northeast of the court-house, that is kept in excellent shape. He has fine horses and carriages, and the terms are moderate. The buildings and outfit are all new, and those who desire a
pleasant drive about this delightful country, with or without a driver,
will here be promptly suited by the accommodating proprietor.
Having noted the leading firms in active business, a brief sum-
mary may aid in giving an idea of the activities that are busy in at-
tending to the varied wants, and promoting the devolopment of this
section. These comprise five general merchandise stores, one clothing,
one hardware and one furniture store, three hotels, three drug stores,
two livery stables, several real estate agencies, a news room, telegraph
office, money-order postoffice, express office, and railroad depot, a skating rink, a weakly newspaper, two billiard rooms, two barber shops,
two millinery and dress-making rooms, one photograph gallery, one
shoemaker's shop, one blacksmith shop, a bakery, a butcher's shop, a
fish market, several restaurants, soda and ice-cream rooms, several
boarding houses, a well-stocked harness shop, a watch repairer, a number of reputable and skillful physicians, from various parts of the
Union, several educated attorneys, several contractors and builders,
two churches with regular preaching, a fire insurance agent, a brass
band, a Masonic lodge, and a variety of societies, agencies, etc., a
court-house and jail, a laundry building, well-stocked with the most ap-
proved machinery, that awaits a capable and energetic manager. A
planing mill, with wood-working machinery, is located in town, and
there are several saw mills in the country adjacent. In fact, quite a
number of industries are located here, but there is room and opportunity for the profitable establishment of many others. Those desiring to better their condition should note these facts.

RETAIL PRICES OF GOODS.

Many who come here are surprised to find the prices so much
more reasonable than they expected, especially in dry goods, clothing,
boots and shoes, canned goods, etc. Crockery and glass ware are
higher than in most of the Northern and Middle States, in consequence
of the high transportation charges on that class of goods, yet they are
not higher than in most parts of the West and South. As regards
provisions: flour is from $6.50 to $8 per barrel, meal and grits $5
per barrel, or 3¼ cents per pound; bacon, 9 to 10 cents per pound;
lard, 12½ cents; hams, 14 to 15 cents; sugar, 6 to 10 cents; rice, 8
to 10 cents; oatmeal, 8 to 12½ cents; crackers, 10 cents; butter, 35
cents; coffee (best Rio), 14 to 20 cents; tea, 50 cents to $1; nails,
4 to 5 cents; beans, 8 cents; syrup, 40 to 50 cents per gallon; kero-
sene, 30 to 35 cents; eggs, 20 to 25 cents per dozen; sweet potatoes,
40 cents per bushel; Irish 50 cents peck; corn, $1 per bushel.

DeLEON MINERAL SPRINGS.

This place of note, that is much visited by the people of Bartow and vicinity, is located about four miles to the southeast of Bartow. Its waters are in great repute as a curative for all diseases of the blood, and of indigestion, dyspeptics finding the use of its waters very beneficial. An analysis proves them to contain the most beneficial ingredients of the most celebrated European medicinal springs. The waters pour forth in immense volumes from unknown depths. The taste is very pleasant. Dr. R. H. Huddleston, the owner, has built a bathing-house here, and many indulge in the luxury of a bath, the spring covering some two acres. A railroad from Bartow to the Springs is in contemplation. Here is also one of the most delightful natural parks, covered with a vigorous growth of trees, to be found in the State. The opportunities for a hotel here are excellent, and some enterprising company can here secure a fortune, as a little expenditure will make it one of the most delightful resorts in America.

Any one having capital which they desire to profitably invest would do well to address Tigner & Tatum, Bartow, Florida, who have full charge of all that pertains to the improvement and disposition of this fine property.

FORT MEADE.

Twelve miles to the south of Bartow is an important centre of business, in a remarkably fertile and delightful locality, that since 1852, at least, has borne the name of Fort Meade. A fort was built here during the Indian war, to keep the uneasy Seminoles in check, and the population has slowly but steadily increased during the past few years. It is doubtless the most fertile and productive section of Florida, but it awaits the near advent of a railroad to give it the necessary transportation facilities and save the twelve miles freightage by team to the terminus of the South Florida Railroad at Bartow. Several roads are chartered that will, without doubt, soon be built through this, one of the most attractive parts of Florida, and then its development cannot fail to make astonishing strides, giving fortunes to those who have been sufficiently far-sighted to invest here.

The busy village is located just west of Peace River, across which the corporation lines extend to the east. Sturdy and vigorous live, water and other oaks give a delightful shade, and thrifty groves of orange trees are very prominent. Lemons, limes, guavas, bananas, etc., flourish and yield very desirable returns. Within the corporation, and within a mile of the post-office, some four hundred acres are set in orange groves, while the trees in nursery can be counted by millions.

Approaching Fort Meade from Bartow, the regular route of travel, by any one of the several roads, you are sure to be much pleased by the attractive beauty of land and landscape. The frequent homes of settlers with luxuriant orange groves ; the numerous cultivated fields, chiefly of corn, pease, sweet potatoes and sugar-cane ; the running streams through the beautiful valleys, the hills and the broad swelling plains, all clothed with a luxuriant vegetation, a dense carpet of grass, magnificent pines, sturdy live oaks, water oaks, with their wide

branching and dense shade, under which the children can pass many
a pleasant hour shielded from the rays of a semi-tropical sun; thrifty
post, willow or turkey oaks, haw bushes that have become trees, wild
plum, cherry and persimmon; though the chief growth is pine, with
scattering oaks, except along the water courses; all give rise to emo-
tions of pleasure.

The streams all flow into Peace River to the east. Along their
banks are dense growths of sweet and black-gum trees, maples, cypress, hickory, live oaks, linden, red bay, cabbage, blue, saw and
needle palm, magnolia, whitewood, ash, iron wood, wild sour orange,
and other varieties too numerous to mention. Here, too, are wild
grapes and running vines that climb the highest trees, too great in
number and variety to be enumerated. Surely here is the field for a
botanist—for a true lover of nature. Here such can devote their time
to study and delightful and refreshing observation.

We must not pause by the way, however, but note as we pass
along, that here, as in other parts of Florida, the land is formed in
strips, or sections, of varying quality. That near Peace River, or on
either side of the creeks that flow into it, is usually of the best quality. The quality of the soil is evidenced by the character of the
varied growths.

Having arrived at the edge of Fort Meade, the attention is ar-
rested by the number, the vigor and the beauty, of the orange groves
grown without the aid of commercial fertilizers. The vigorous water
oaks along the streets and in the yards, also demand and receive our
admiration. The lovely carpet of Bermuda, or other grasses, with
which the streets and fields are covered, is suggestive of fine, fat cattle
and plenty of milk. Bees, too, would evidently do well here, hence
it should be "a land of milk and honey."

A knoll to the northeast of the village, whereon is a house, beneath the shade of majestic oaks, and adjoining an orange grove, is
pointed out as the site of the old fort   The Methodist Church is to
the north of the business centre, and the fine new two-story school
building is south of west. Across Peace River, to the east, is a long
bridge. The town plat is surveyed into lots of four acres, separated
by wide streets that cross each other at right-angles. The business
houses are located on Main Street, the post office and drug store on
the corner of this street and Orange Avenue, being the centre. To
the east is the Adams House and to the west is French's Hotel.

Observation shows that there are four general merchandise stores,
two drug stores, two hotels, two pool-rooms, a post office and a telegraph office, a millinery store, a barber shop, a livery stable, a blacksmith shop and tools awaiting an enterprising man to put it in operation, a public library, a number of real-estate offices, several physicians, and last, but not least, a live newspaper, the Fort Meade *Pioneer*.
To the north is the Methodist Church, to the west the large two-story
school house. The residences are scattered about in country style,
the most of the houses and the 20,000 orange trees (about 6,000 in
bearing) being within a radius of one mile of the post-office. The
population is about three hundred. The voting precinct, of which
this is the centre, polled 214 votes at the last election.

From the earliest settlement, Fort Meade has been a trading centre of considerable importance, it being the supply point for a large extent of country, especially to the east and south. Here the traders met the drovers from the outlying ranges; here many a sale and exchange has been made; here for many years has been the camping ground.

Fort Meade was incorporated March 16th of the present year, 1885. It is located in the southeast corner of Township 31 south. Range 27 east. Being four miles square, it covers an area of sixteen square miles of the most fertile, productive and attractive lands in the State. The business part of the corporation is in Sections 26 and 27, Main Street being just north of the dividing line. John Jackson, Deputy Government Surveyor, ran the exterior township lines in 1854, and it was sectioned by W. G. Mosley in the following year. From the *Field Notes* I quote the following very unusual and very flattering :

"GENERAL OBSERVATIONS.

"This township is finely adapted for agricultural pursuits, the land being mostly of first and second rate quality pine, with dark brown rocky soil and undulating surface. The Tallakhchopka River, or Pease Creek, runs through it from north to south, with a narrow stream and flat banks, and wide, thick swamps subject to overflow from freshets.

"The same deadening extends through it north and south and from two to three miles in width. The western tiers of sections are flat pine and ponds, third rate land. Settlements thick and numerous throughout the whole township."

The early settlers invariably selected the most fertile and productive lands, for they were compelled to secure their subsistence from the soil ; hence, as large crops could be raised here without commercial fertilizers, the scenery also being very pleasing and attractive, and the climate delightful, it is no wonder that the residents, and also the increasing numbers of visitors, should deem it the finest section of Florida. Its low latitude gives it great advantages in the raising of citrus and other semi-tropical fruits, frosts and freezes being very rare. The thrifty orange groves show the excellent adaptation of both soil and climate. Here vegetables grow with wonderful luxuriance all the year, well rewarding the cultivator's attention. The town has two telegraph lines and a daily mail, and will doubtless have a railroad within a year, surveys having already been made.

A stopping place is necessary for the prospector while he is determining where to locate. In this respect Fort Meade is fortunate, having two pleasant hotels. The ADAMS HOUSE, just east of the post-office in the centre of a four acre orange grove, and near the beautiful live oaks and delightful scenery that adorn the river's banks, is a very pleasant and convenient stopping place. Mr. A. H. Adams, with his agreeable family, from Seymour, Ind., is the obliging and attentive landlord.

V. L. TILLIS finds himself equal to the task of running the post-office, a drug store and two telegraph lines, though he has to get around lively at times. In fact, he is always busy.

MRS. EDNA HAYMAN has a very neat and quite attractive millinery store, and is emphatically a woman of business, as well as polite, agreeable and entertaining. She naturally has hosts of friends and, we understand, gives excellent satisfaction.

HENDRY & CARTER, dealers in general merchandise, are located on the corner opposite the post-office and are doing an immense business. They are active young men, and their store and warehouse is literally packed with goods of every variety and description in general use, as well as a great variety of miscellaneous articles, their endeavor being to supply every demand.

J. N. HOOKER & Co. are located to the west, and appear to be doing their full share of the trade. They have a large and well-assorted stock of general merchandise, suited to the needs of the country. This store was opened previous to the one at Bartow, and Colonel Hooker makes it frequent visits, though it is under the management of able and trusty clerks.

The LIVERY STABLE of Wilson & McKinney next attracts our attention, being a great convenience, as well as necessity. The building is commodious and well-arranged for the large number of horses and carriages that are kept on hand, for sale, or for the benefit of the travelling public.

The FRENCH HOUSE, just beyond the livery stable, is situated about half way between the post-office and the fine, new school-house. J. L. Bettis, the genial landlord, was recently from Jacksonville, and has a wide acquaintance and an extended knowledge of the .country. The rooms are pleasant and the table attractive, while the quality of the cooking, etc., is of exceptional excellence.

C. C. WILSON, the practicing attorney, has a very pleasant residence half a mile west of the post-office, where a new centre is being established. Though comparatively young, he is a representative man, being the Delegate-at-large for Polk and Manatee Counties to the Constitutional Convention. He has several promising groves, a great variety of choice and rare fruits, is a practical experimenter, and is largely interested in the lands of this section.

The SUNNYSIDE NURSERIES of Mitchell & Hester, to the extreme west, with their choice varieties of oranges, lemons, limes, plums, persimmons, figs, grapes, peaches, mulberries, roses, cedars, arbor vitæs, etc., besides a few each of plants too numerous to mention, must be seen to be fully appreciated.

Capt. F. A. WHITEHEAD, one of Fort Meade's leading and most active and influential citizens, has a delightful residence amid towering oaks and a fruitful orange grove, in the heart of the village. He also has a variety of pleasing growths, such as Japanese plums and persimmons, Peen-To peaches, lemons, limes, pine-apples, strawberries, bananas, mangos, sapadillos, grapes of numerous kinds, flowers in great variety, and other things too numerous to mention. He also has a large number of acres of the choicest citrus fruits in grove. A native of New York City, he has made good use of his thirteen years in Florida. Resigning his position in the navy at the close of the war, he made a thorough acquaintance of California, and has been in the fruit and stock business ever since, yet having a farm in Delaware. He prefers Florida to any State, has large tracts of land here and is doing a very extensive real-estate business.

E. E. SKIPPER is an extensive dealer in lands, knows the country thoroughly, and can suit every taste or condition, as he has every variety and price, both unimproved and improved. His faith in the country is shown by the fact that he has some 3,000 trees in grove, of which about 400 are bearing, some being from twelve to fifteen years old.

R. C. LANGFORD has a very pleasant and productive place in a pine and oak clearing about a mile southwest of the postoffice, with which he is connected by a private telegraph wire. Here he has a superabundance of fruits, vegetables and other farm and garden products that would astonish those who think nothing can be produced in Florida. He raises them in his grove year after year. He has choice tracts of land all over the country, and makes a business of buying and selling lands.

J. E. ROBESON, a practical surveyor and dealer in lands, has been thoroughly identified with the interests of Fort Meade since 1872, the past eight years being chiefly devoted to surveying and selecting lands. He graded large quantities of the Disston and also of the Sir Edward Reed lands, and is now devoting his time and knowledge to the benefit of the public who are so fortunate as to secure his services.

DR. M. O. ARNOLD, recently from southeastern Iowa, charmed by the attractive beauty of the country, located at Fort Meade. Finding the country very healthful and desiring a broader field, having had five years practice in his profession, he has become interested in aiding others to secure homes in this delicious land, and many are being benefitted by his efforts. He was formerly treasurer of the South Florida Land Company, of which Dr. C. C. Mitchell, the present State Commissioner of Lands and Immigration, was president. He is now agent for the Florida Land and Improvement Company; The Kissimmee Land Company; The Atlantic and Gulf Coast Canal and Okeechobee Land Company, and The Florida Land and Mortgage Company (limited). He gives special attention to tracts for colonization, and to town plats for settlement, both large and small. He has a tract of 12,000 acres, suitable for towns or colonies, for a nominal figure; also, several miles of gulf frontage, in a tropical climate, with some very fertile lands, as well as large tracts with fine natural grasses, especially suitable for stock ranges. Besides these heavy and desirable outlying lands, he has extensive interests at Fort Meade and vicinity.

DR. C. F. MARSH, recently from Mount Pleasant, Iowa, has a high reputation as a skillful practitioner.

DR. J. WEEMS, formerly of Missouri, is also a pleasant and capable physician, ranking high in the profession.

BLACK & EDWARDS, real estate agents and civil engineers—J. F. Black, of Illinois, and J. A. Edwards, of Alabama—have a good line of grove property, town lots and wild lands. They buy and sell on commission, give careful attention to surveys and titles, and give all possible assistance to those who desire to better health or fortune by locating on the fertile lands in the delicious climate of Polk County.

JAMES WYNN, a competent builder and contractor, is about to establish the saw-mill, which he has purchased, convenient to the town, and will furnish lumber and erect buildings at favorable prices.

PHILIP DZIALYNSKI has, for a number of years, been prominently identified with such interests as tended to the development of the town, and largely interested in its affairs, during the several stages of its growth.

G. W. HENDRY, who became a resident of Fort Meade in 1852, being then a stout boy, has written and published an interesting descriptive pamphlet of Polk County. When he came, this section was occupied by a company of troops at the fort, but there were no settlers, unless his elder brother, F. A. Hendry and family, with his father-in-law, Louis Lanier and family, who had the first herds of cattle driven east of Peace River, and were engaged in supplying the soldiers with beef, be so considered. Mr. Hendry is and has been actively engaged in locating land, having a thorough acquaintance with all South Florida, and unquestionable authority.

THE DEADENING.

A great natural curiosity, called the "The Deadening" exists at and about Fort Meade, covering a tract of country some ten miles or more from north to south, and some five or six miles from east to west, being divided by Peace River. When the first settlers came, in the fifties, they found the whole tract entirely divested of living trees, except along the water courses and on the higher knolls. Lying prone on the ground or standing erect, like neglected and forsaken sentinels, were the solid remains of what had years before been a vigorous growth of pine.

The cause of this destruction of the trees is utterly unknown. Various theories have been adduced, but none are fully satisfactory. G. W. Hendry claims hail to have been the agent of destruction, but this theory is untenable from the fact that no hail-storm was ever known to cover such an extent of territory, and besides hail-storms are unknown here. It will also be noted that the trees on the highest knolls along the water courses, and in the lower lands were untouched. Others claim high water to have been the cause. The most probable explanation is that several wet seasons prevented the usual forest fires, permitting the dead grass and leaves of the trees' to accumulate in great abundance. Then came a very dry season, fire raged throughout the forest, and its intense heat killed the trees. Whatever the cause, the country assumed the appearance of the Western prairies. Since the first settlement, vigorous growths of oak and pine are springing up over the whole area of these, the choicest of lands, and were it not for the rapidly increasing settlements and groves it would soon be forest again.

LAKELAND.

One of the radiant gems of South Florida's many new yet rapidly developing towns is Lakeland, the growth of a little more than a year. It has a fresh, thrifty, prosperous and substantial appearance, while in every direction are made manifest the evidences of resolute vigor and determination. The face of the country is beautifully rolling, carpeted with thrifty grasses and covered with vigorous growths of oak, pine, etc. At frequent intervals the surface is indented with bright jewels in the shape of lovely clear-water lakes of varied form and size, wherein are mirrored the beauties of nature; the form and foliage of a thrifty and delightful vegetation, the fleeting clouds, the twinkling stars, the soft radiance of Luna, night's resplendent queen, or the bright effulgence of Old Sol, the glorious king of day.

Lakeland is situated near the central part of Peninsular Florida, and of the lovely and attractive County of Polk, as well as in the highest, most healthful and delightful portion. The railroad survey gives it an elevation of 210 to 217 feet above the sea level. Here is found a great and pleasing variety of scenery, some of the deep, clear-water lakes with their clean, hard, grass-covered banks, being from forty to sixty feet below the higher points of the plateau. They are not grassy ponds, but pure, deep-water lakes, whose banks afford the most delightful and healthful of sites for residences, for lovely homes, and they are being appropriated quite rapidly. There are nine of these attractive sheets of water within a radius of a mile of the town, almost entirely free from mud and marsh, and abounding in fish, giving delightful opportunities for recreation, as they are situated in every direction from the centre. They also give delightful views, and the air passing over them is imbued with an inspiriting freshness.

The surface soil is varied, none being below the average, while a peculiar feature of the soil on some of the elevations is that it is almost as rich as hammock, and preferable for many reasons. Fruits and vegetables thrive and yield magnificent returns. At a depth of two to eight feet, and outcropping at some places, is a sub-soil of yellow clay. The water is excellent.

The beautiful forests are fast disappearing and in their place are scores of handsome and substantial buildings, thrifty groves and cultivated fields. Everywhere is heard the ceaseless hum of busy industry, transforming the face of nature. The South Florida Railroad passes through the incipient city from east to west and the Florida Southern coming from the north here forms a junction with it. The expectation is that it will soon be extended to Charlotte Harbor on the south.

Section 18, Township 28 south, Range 24 east, is the centre of the corporation, which also embraces portions of Sections 7 and 19 in the same Township and Range, all of Section 13 and parts of 12 and 24 in Township 28, Range 23, thus embracing two whole sections and parts of four others, and that, too, in one of the most delightful, agreeable and satisfactory parts of Florida, as regards deliciousness of climate, healthfulness of location, excellence of water, freedom from insect and other pests, general fertility and productiveness of soil, exemption from destructive frosts and freezes, genial breezes and salubrity of atmosphere, excellence of society, active, enthusiastic and vigorous energy of the rapidly-increasing population, handsome and substantial character of business edifices and private residences, ease and facility of communication with other parts, by railroads, telegraph, etc., and numerous other attractions that will suggest themselves to the visitor.

Lakeland is regularly laid out, with broad streets crossing each other at right-angles. In the centre is a park of three acres, that is to be adorned with trees, shrubbery, etc. To the north of this is the elegant depot of the South Florida Railroad. Around this double square, the town, which was incorporated January 1st, 1885, is rapidly assuming an undeniable substantiability, about two hundred buildings having already been constructed, while more are under contract, yet in February, 1884, there was only one rough frame building and two log shanties for the railroad hands. Now there are several fine hotels, numerous general merchandise stores, hardware, feed and drug stores, restaurants, boarding houses, pool-rooms, express, telegraph and post-offices, saw and planing mills, shoemaker's shop, livery stable, millinery, gent's and ladies' furnishing goods, real estate and other offices, in fact the usual variety of avocations of some six hundred inhabitants. Also, well conducted schools, churches, etc., and a wide-awake newspaper, the Lakeland *News*, L. M. Ballard, Editor and Proprietor. He is also the proprietor of the North-Side Hotel.

Prominent among the real estate agents, with handsome and convenient offices centrally located, are Green & Munn, Torrence & Bristow, Scott & Roquemore, who will furnish all desired information regarding lands in this vicinity and other parts of South Florida.

NEWMAN & Co. have a pleasant store and a fine stock of gent's and ladies' furnishing goods, boots and shoes, notions, etc.

O. J. FRIER has an extensive and well selected stock of general merchandise, at satisfactory prices.

W. B. BONAKER, dealer in general merchandise, endeavors to meet every demand in that direction, at prices to suit.

S. L. & H. J. DRANE, druggists and apothecaries, are well prepared to fill any demands in their line.

Society is decidely intellectual and progressive here, as is shown by the excellent schools, the several religious and other societies, the Methodists, Baptists and Presbyterians all having organized and energetic societies, while the schools are well sustained, and very prosperous under the management of capable teachers.

Town lots sell for from $50 to $1,000, according to size and location, while outlying lands are from $2.50 to $100 per acre, according to distance from the centre, quality, and desirability of location.

The vigorous growth and advantages of Lakeland and Polk County have been well shown by the enterprising real-estate agents, Torrence & Bristow, who last year published an excellent pamphlet that had a wide circulation, and gave much desired information.

ACTON.

This progressive and thriving headquarters of enterprising Englishmen owes its existence to the energy of Piers Elliot Warburton, Esq., who formerly held the honorable rank of Lieutenant in the English navy. Mr. R. W. Hanbury, an Englishman of large estates and income, is the largest property owner. This is to be an English headquarters. Here is located the Florida Mortgage and Investment Company, limited, Mr. Warburton, being the manager, that has unlimited amounts to loan on security, at reasonable rates of interest.

Acton is located on the South Florida Railroad, half way between Kissimmee and Tampa, and one mile east of Lakeland. The town site is quite level, extending between Lakes Bonnie and Parker, two very attractive sheets of water. The streets run from east to west and are crossed by avenues from north to south. Here are very complete saw and planing mills, a hotel and several boarding houses, general merchandise stores, a $2,500 school-house, a post-office and depot, a real estate agency, Mr. Warburton's loan office, and quite a number of pleasant and attractive dwellings, and contracts made for a number of other elegant and substantial structures, the terms offered being very favorable.

C. H. ALLEYNE & Co. (limited) have a very fine new office opposite the depot, do a very extensive real estate business, and furnish and desired information respecting Acton and other parts of South Florida.

HAINES CITY.

The vigor of the growth of Polk County is well shown by the numbers of busy towns that are springing into existence, in the most favorable and delightful of locations and now that this section is well provided with railroads, which are being rapidly extended, its development is one to greatly exceed anything heretofore seen in Florida.

Haines City is no chance growth, but the result of the deliberate premeditation of several of the most active and far-seeing men, whose influence is felt throughout the State in its development. It is located on the South Florida Railroad, about half way between Sanford and Tampa, the town plat occupying the east half of Section 29, Township 27, Range 27, it being in the beautiful Lake Region of Polk County. The elevation of the railroad bed here is 210 feet, which is claimed to be the highest on the line from Sanford to Tampa. The country is rolling and interspersed with beautiful lakes and airy, delightful elevations, from forty to sixty feet above their pure crystal waters, affording very favorable sites for sanitariums, residences, etc.

The prevalent growth is pine, interspersed with large quantities of live, water, willow, post and other oaks, while along the lake shores are frequent hickories. India rubber and paw-paw trees are also found growing wild, proving conclusively that severe frosts or freezes have had no place here. The ground is covered with a dense growth of grass, the timber is very thrifty, and the soil in many places is of a chocolate color, underlaid with the yellow subsoil so necessary to the best development of the orange and other citrus fruits. It is underlaid with clay. In fact, all kinds of vegetation thrive here.

The town was laid out last January, by Frank J. Hinson, a man of thorough experience, who is now the resident agent and manager. Harrison Jones, who has had eighteen years experience in four of the counties of Florida, also has extensive interests here, where he has made his home, greatly prefering it to all other localities.

Haines City is building up very rapidly, the advantages of the location becoming readily apparent to any one who will take the trouble to investigate. Lands can be secured at very reasonable prices, and are being taken quite rapidly.

Though so short a time has elapsed since the survey, there is already a post-office, a hotel, a saw and planing mill, general merchandise stores, boarding houses and elegant cottages. Building and clearing is the order of the day. Quite a number of families have already located in this desirable and healthful location, and a public school is to be opened in the autumn. The opportunities for boating, fishing, hunting, gardening, or making a fortune, are most excellent, while the lake views are delightful.

In addition to the incorporated towns mentioned, there are many localities throughout Polk County where a beginning has been made, as well as numerous others yet unheard of, that will no doubt soon assume importance. Each section has its own special advantages that can be best determined by personal observation. The people of Polk County are content, feeling assured that of all localities they possess the most superior advantages.

POLK COUNTY'S FERTILE RIDGE.

A RIDE THROUGH THE BEAUTIFUL COUNTRY ALONG PEACE CREEK.

[Special Correspondence *Florida Times-Union*.]

FORT MEADE, June 20, 1885.

Thinking that a few words regarding this section of rapidly-developing South Florida might be acceptable to the thousands of your readers in various parts of the Union, I contribute my mite for their benefit. Your valuable journal is not only a recognized authority in regard to matters in all parts of the State, but it is also a very important factor in aid of the remarkable progress that is being made. It is, in fact, a necessity to all who desire correct knowledge regarding the capabilities, advantages and progress of the varied sections of this great State, destined, ere long, to be one of the most wealthy and prosperous States in the galaxy of the Union. This is to be a land of lovely and attractive homes, as well as the chief resort of the invalid and tourist.

Leaving the busy cars of the South Florida Railroad, and bidding adieu to Captain Badeau, the genial and accommodating conductor of the branch road, at Bartow, I take a look about the pleasant and fast-growing town. Its streets are wide and cross each other at right-angles. Numerous new buildings, completed or in progress, are seen in every direction. The fine court-house, the most attractive and commodious in South Florida, arrests our attention. It is situated on a commanding rise of ground, the centre of the business portion of the progressive town. The square acre that surrounds it is inclosed with a row of live-oak posts, painted red, through which cable chains are run. Sixteen feet within is a neat picket fence, painted white. The court-house roof, with its four gables, has just been covered with cypress shingles and painted a dark red.

But the most attractive feature of the town, not excepting the indications of solid progress, evidenced by the numerous new buildings, is to be found in the beautiful oaks, that greet the view in every direction, and afford such delicious shade. They even enhance the feeling of sure solidity that is derived from the firm tread of the ground, which is quite in contrast with many other sections.

But it is dinner time, and at half past one, P. M., the demands of the inner man override, if not suppress, the desire to indulge in contemplations of the beautiful. An abundant and toothsome meal, wherein home-grown vegetables play an important part, neatly-served and well-cooked, is secured at the Bartow House, and I devote the balance of the day to observation and reflection on the many advantages that this section affords to enterprising men from all sections of the Union, and especially to men of moderate means with families.

Here I find an extensive tract of fertile pine and oak lands, and learn that the settlers have been self-supporting from the very first. The South Florida Railroad now has its terminus here, to the southeast, and a survey of the Florida Southern runs through the corporation, just to the west of the centre of this attractive capital of Polk County.

During the evening it was my good fortune to make the acquaintance of Dr. C. C. Mitchell, a distinguished resident of Fort Meade, who has been very appropriately appointed as Commissioner of Land and Immigration, by our able and clear-headed Governor, General Perry. The result of our interchange of ideas was an earnest and courteous invitation to visit that noted section of balmy Florida, of which I had heard much but had seen nothing. Cancelling some other engagements, I cheerfully accepted the proposition, and early the next morning we were whirling rapidly to the southward.

The genial doctor is a good judge of horse flesh, and drives an excellent team. The country through which we sped was a surprise, it was so different from many other sections that I had visited. The roads were hard and firm, and as easily traveled as those at the North. There was an absence of deep sand and of annoying dust. New buildings, some of them of an elegant character, were seen on either hand; also, many a beautiful orange grove, whose thrifty growth and exceptionally dark green leaves, betokened a fertile and productive soil. Promising fields of vigorous corn are quite numerous, indicating that the people are inclined to raise their own supplies, and not put all their trust in the orange crop, not at least until a railroad should be extended from Bartow or Lakeland, to give them better facilities for transportation and ready access to Northern markets.

The general aspect of the country was very pleasing and attractive, making a very satisfactory impression on the mind of the visitor. It may be described as a broad plateau of fertile and productive lands, extending some three miles west of Peace River, from Bartow to Fort Meade, and a few miles beyond to the north and to the south. The surface is generally undulating in broad swells, with here and there a handsome knoll that would furnish an exceptionally pleasant and salubrious building site. Numbers of them are so occupied, and pleasant homes with luxuriant groves of orange trees, laden with abundant promise of the golden fruit, as well as varied farm crops, the most notable of which are thrifty corn, pease and sugar-cane, occupy occasional clearings.

The forest growth, away from the river bank, is chiefly pine, interspersed here and there with wide-branching live oaks that, with the firm tread of the ground, gives an impression of substantial and enduring stability. There are also many water oaks, whose thrifty and vigorous growths give a delicious shade that is highly appreciated, especially by visitors from the North and West. Post or willow oaks are also quite numerous, and attain a greater size than in many other parts of Florida. Here, too, the haw becomes a handsome tree, instead of a bush by the wayside. The wild persimmon is also abundant.

As the genial and thoroughly-informed doctor and myself speed over the country, by the west road, known as Broadway, as we leave Bartow, we make occasional detours to the right or to the left, either bodily or mentally, by the doctor's intelligent and far-reaching descriptions, intermingled with scraps of history or personal adventure.

The population is very much scattered, every settler evidently endeavoring to secure all the elbow room possible, that there might be no danger of conflicting interests and consequent animosity. Though since the first settlement of this section, in the fifties, some thirty or more years since, the people have raised considerable quantities of corn, pease, sweet potatoes, rice and sugar-cane for domestic use, as well as some cotton for shipment their chief wealth has been in their fine herds of cattle and countless swine, which are here of much more pleasing form and quality than in the tier of counties to the north. Here, the shafts of wit leveled at the "razor-backs," the "pine-rooters," the destructive and remorseless vagrants of other sections that have all the worst characteristics of the creatures by courtesy called "hogs," have no place. They appear to be of good form. sleek and fat, and evidently do not have to root very hard or persistently for a living. The hog here, undoubtedly, has found his paradise, the abundance of mast from the numerous oaks and the esculent roots, found in the lower lands and along the courses of the numerous running streams that wend their way to Peace River, supplying them with an abundance of nutritious food.

The water courses are quite numerous, flowing from the flat woods of the west, and serving as natural drains to the section between them and Peace River. These creeks break the plateau along the river into ridges and give a great and pleasing variety to land and landscape, which give an attractive homeland character to people from the North, making them feel much more at home than is possible in unbroken tracts of strictly pine country.

In valleys along the banks of these creeks, and beyond, are magnificent and enduring live oaks, choice sweet-gums, majestic cypress, cabbage palm, beautiful water oaks, attractive maples, wild sour orange, black-gums, turkey oaks, tall and sturdy hickories, sweet bays, magnolias, white-wood, haw, persimmon, abundance of beautiful pines, wild cherry, and quite a number of other forest growths. There are also numerous vigorous and thrifty climbing vines and creepers, a great variety of shrubs, wild plants, weeds, etc.; in fact the natural productions of all the zones, except the frigid, seem to have centered here. The country is very pleasing, but quite unlike either the pine or hammock sections of other parts of Florida, and people from all portions of the country can here find particular attractions. The slopes of the valleys gave me especial pleasure.

Reaching Dr. Mitchell's pleasant residence, on the west of the village, we found his men busy in the branches of a wide-spreading oak that overhung the yard, brushing a swarm of bees off a large limb, and cooling down their aggressiveness of disposition with a plentiful supply of fresh water. But an excellent dinner was ready for serving, and I will defer remarks upon his fine groves of orange trees and his many acres of nursery, wherein are growing not only orange and lemon trees, ready for transplanting, but also thousands of roses and a great variety of small fruits, shrubs, grasses, etc. In fact, I understand that the intention is to grow every species that may prove to be desirable or useful, and I shall watch the progress of the experiments with great interest, as it may be the means of adding many thousands of dollars to the value of the annual productions of this delightful land. "SHERMAN."

HISTORY OF ORANGE CULTURE IN FLORIDA.

BY REV. T. W. MOORE, D. D.

Some time before the discovery of America, the sour orange—the brigerade—was introduced into Italy, and a short time thereafter it was carried to Spain. The Spaniards brought this variety to Florida. The sweet orange was then unknown in Europe. Doubtless the Spanish Catholic missionaries first distributed the seed of the brigerade—frequently called the Seville—orange in the vicinities of the Spanish forts and missions. As the fruit multiplied, the seeds were scattered by the Indians along the banks of the rivers, near their camping grounds, usually points projecting into the rivers. Thence they were scattered throughout the State of Florida.

The largest of those wild orange groves, twenty and fifty years ago, were found along the eastern and southern shores of rivers and lakes, and in the hammock and swamp lands of Florida. In addition to the protection from damage by the frosts to the young plants afforded by the water, the hammock and swamp lands gave protection against fires, which annually swept over the pine woods, destroying the slow-growing trees. Some of these wild groves were, fifty years ago, cut down and the land cleared for planting corn, cotton and cane. This was repeated as late as twenty-five years ago, before the monetary value of the orange was appreciated in this country.

One hundred years after America was discovered, the sweet orange was introduced into Europe. Later it was brought to Florida, and a few trees were planted in St. Augustine, and afterwards in the settlements along the St. Johns and Indian Rivers. The pollen of the sweet orange fertilizing the flowers of the sour, produced the hybrid "bitter-sweet." At the close of the civil war small plantations of sweet oranges were found throughout the State; consisting usually of a few trees growing around dwellings. There were a few groves of larger size, ranging from four hundred trees to nine hundred, in the vicinity of St. Augustine and along the St. Johns River. The largest in the State was planted by Dr. Speer, at Fort Reed, near Mellonville, and the Dummitt grove on Indian River.

About the time Dr. Speer planted his grove quite an interest in orange growing sprang up in Florida and many groves were planted along the banks of the St. Johns. But in an evil hour fresh plants of the orange from China were introduced and planted at Mandarin. They were infected with the scale insect. The trees in the vicinity of Mandarin were the first to be destroyed by the insect. At that time the hundred and one natural enemies of the scale insect had not come to the rescue of the orange grower as now; besides, the orange grower of that time did not know of modern appliances and remedies. The scale spread from grove to grove, and in a short time sweet and sour orange trees yielded to the invading host of the foreign enemy. The frost of 1835 having cut down the trees, from the effect of which the old trees were beginning to recover when the scale commenced its ravages, combined to produce the impression among the old settlers that the orange prospect was forever blasted.

At the close of the war, many of the old trees, both sweet and wild, had recovered from the effects of both insects and frost, and were bearing liberal crops of such fruit as travelers from all parts of the world had never before eaten. The fruit sold at good prices. Some of those who had lately come into the State thought there was a living in an orange grove. Land was bought and planted in wild sour stumps. Seed beds were planted for nursery stock and acres were set with young plants. We were told that by the time our trees were ready to bear we would be in another country where there would be no need of planting. We answered, then we would plant for our children. We were told that by the time the trees were in full bearing oranges would not be worth picking in Florida. Though some of us were threatened with the lunatic asylum, we still persisted in planting and cultivating the orange. The evil prophecy failed. Other persons caught the orange fever, until finally the old prophets were converted and are to-day our most enthusiastic orange growers. To-day hundreds of thousands of trees are growing, and tens of thousands more of plants are ready to be set in groves.

WILL THE BUSINESS BE OVERDONE?

The question now comes up, will not the business be overdone? We answer no. With the small area within the United States capable of producing oranges this will be impossible. Canada and the United States are rapidly increasing in population and these alone could consume the entire product from the orange-growing sections of the United States. But the Florida orange is the finest grown and will ultimately command the markets, of Europe as well as America.

Occasionally already a glut in the market has occurred, but this has been in each instance the result of (mainly) a double fault of the producers. They have attempted to narrow the marketing season to three or four months, when it should be extended over from eight to twelve months. Oranges will remain on the trees in good condition six months after they have turned yellow. Properly handled and cured they will keep several months after they have been clipped. The Florida season for marketing, like the European, should embrace the entire year. The second mistake to which allusion is made was the result of the destructive hurry peculiar to Americans. The fruit was gathered green, carelessly handled, packed without being properly cased, much of it infested with fungi and then gathered, packed and shipped, through all sorts of weather. Such fruit rapidly spoiled. Careless handling of transportation companies added to the disaster, and hence the merchants had to sell what sound fruit might reach them at low prices or throw it away.

Orange culture will pay beyond any other agricultural pursuit, even should the price fall to 75 cents per box. When reduced to that price fifty million boxes would not over-supply the present population of the United States and Canada. There are thirty States producing apples and peaches, and yet both these crops, which have to be marketed within a few weeks or months, are grown with profit. With such facts before us, we have no fear as to the over-production of the orange.

A FASCINATING VOCATION.

To those engaged in the business, orange growing is truly fascinating. The beauty of the tree, the beauty and fragrance of the flower, challenge all rivalry among ornamental trees and beautiful flowers. The æsthetic cultivator becomes a true lover of his sweet and beautiful pet, which he looks upon as a relic and reminder of paradise. But when this beauty is accompanied with useful, golden and gold-bearing fruit, affording a living, and promising all other material luxuries, then the lover appreciates his orange grove only less than he appreciates his wife, who has brought to him not only the accomplishments of a sweet and cultivated woman, but with herself an ample fortune. And though he may have waited as long as Jacob did for his Rachel, he does not regret the toil and waiting since the reward is ample. I do not know but that the toil and waiting demanded by the orange does not increase the ardor of the planter, and increase his pleasure when once the tree has been brought to full beauty and bearing, for we love best those that need to be courted earnestly in order to be won. When thus won we feel that the bride is the more fully our own.

HOW TO GROW THE ORANGE.

Does the reader wish to know how to win this fair bride, clad in nature's richest green, adorned with golden globes, crowned with fragrant orange blossoms—her own fair crown, so often plucked for other bridal wreaths? Did space permit in this full sheet of the *Times-Union*, further writing would not be necessary, for are not all these things written in the books of the chronicles of many writers on "Orange Culture" from Maine to Texas? These have all written you about the seed-bed, the nursery, the planting, suitable locations, the gathering and the shipping.

THE QUANTITY OF ORANGES SHIPPED

the past season was about six hundred thousand boxes; the present year the crop may reach a million boxes. The crop of 1868 only reached a few thousand packages, and had so slow a sale that it had to be extended to as late as May to find buyers. The price prevailing at that time was $7.50 per thousand. The price has gone up with the production. During next May, if they can be found outside of New York, the Florida orange will sell for not less than $4.50 to $5 per box.

THE FLORIDA TRADE MARK,

The excellence of the Florida orange is now so generally known that many other oranges are sold under that name. The writer knows no way to avoid this imposition except to stamp each orange grown in Florida with the inimitable Florida trade mark. No other country has yet produced the russet. The brown tinge mars the beautiful golden color, but it makes the orange bearing this stamp all the sweeter, and, like Cæsar's wife, above suspicion. Nature has thus given us an impost protection against foreign competition, which the Government cannot take off. What goddess or nymph was it that covered herself with soil to save herself from violence? She was the sweeter and safer because of her soiled exterior. So with the orange. The dingy russet is best.

—*Exposition Number Florida Times-Union.*

www.ingramcontent.com/pod-product-compliance
Lightning Source LLC
Chambersburg PA
CBHW021958190326
41519CB00009B/1307